AF838954

Imperial Military Transportation in British Asia

Also available from Bloomsbury:

The Collapse of British Rule in Burma: The Civilian Evacuation and Independence, by Michael D. Leigh
Burma 1942: The Road from Rangoon to Mandalay, by Alan Warren
Military Transition in Early Modern Asia, 1400–1750: Cavalry, Guns, Government and Ships, by Kaushik Roy

Imperial Military Transportation in British Asia

Burma 1941–1942

Michael W. Charney

BLOOMSBURY ACADEMIC
LONDON • NEW YORK • OXFORD • NEW DELHI • SYDNEY

BLOOMSBURY ACADEMIC
Bloomsbury Publishing Plc
50 Bedford Square, London, WC1B 3DP, UK
1385 Broadway, New York, NY 10018, USA

BLOOMSBURY, BLOOMSBURY ACADEMIC and the Diana logo are trademarks of Bloomsbury Publishing Plc

First published in Great Britain 2019

Copyright © Michael W. Charney, 2019

Michael W. Charney has asserted his right under the Copyright, Designs and Patents Act, 1988, to be identified as Author of this work.

For legal purposes the Acknowledgements on p. ix constitute an extension of this copyright page.

Cover image © Railway in Burma, 1911. From the author's personal collection.

All rights reserved. No part of this publication may be reproduced or transmitted in any form or by any means, electronic or mechanical, including photocopying, recording, or any information storage or retrieval system, without prior permission in writing from the publishers.

Bloomsbury Publishing Plc does not have any control over, or responsibility for, any third-party websites referred to or in this book. All internet addresses given in this book were correct at the time of going to press. The author and publisher regret any inconvenience caused if addresses have changed or sites have ceased to exist, but can accept no responsibility for any such changes.

A catalogue record for this book is available from the British Library.

A catalog record for this book is available from the Library of Congress.

ISBN: HB: 978-1-3500-8945-7
ePDF: 978-1-3500-8946-4
eBook: 978-1-3500-8947-1

Typeset by Newgen KnowledgeWorks Pvt. Ltd., Chennai, India

To find out more about our authors and books visit www.bloomsbury.com and sign up for our newsletters.

Dedicated to my niece and nephews
Anna, Ryon-Ryon, Sohei and Wayne.

Contents

List of Illustrations viii
Acknowledgements ix
List of Abbreviations xi

Introduction 1
1 Metropolitan Transportation Technique from Britain to India and Iraq 15
2 Local, colonial railway experience 33
3 The failure of civilian management of a colonial railway in wartime 51
4 Militarization and the Rangoon to Prome Branch 73
5 Kings of the road: Decentralization and local initiative on the railway 97
6 The technical limits of the railway: The Mandalay to Lashio Branch 133
7 Dark territory: The Myitkyina Branch, late April to early May 1942 147
8 After Burma: The militarization of railways in India 157
Conclusion 181

Notes 187
Bibliography 213
Index 227

Illustrations

Figures

7.1	Ratio of the number of sidings to railway mileage on the Burma Railways during the First Burma Campaign	153
7.2	Locomotive density during the First Burma Campaign	154

Maps

0.1	Burma Railways circa 1941	xiv
3.1	The Burma Railways and the war in the Southeast, January–March 1942	66
4.1	Militarization of Burma Railways, late February 1942	83
5.1	Burma Railways in April 1942	116
6.1	Burma Railways South of the Irrawaddy, end of April 1942	143

Tables

2.1	Racial divisions among Burma Railways staff in 1940	35
2.2	Geographical structure of the Burma Railways administration – HQ of departmental districts, circa 1941	43
2.3	Section and branch line names of the Burma Railways, circa 1941	44

Acknowledgements

Over the course of the last few years, many people contributed to this book in one way or another, whether through conversations on the railways, warfare or both, or by reading the various versions of the manuscript or sharing theirs, and it is hoped that they see in the manuscript improvements for which I am extremely grateful. Special thanks is owed, however, to the Institute for Advanced Studies of Asia at the University of Tokyo which hosted me as a project professor for two years, from 2012 to 2014, and to the School of Oriental and African Studies (SOAS), the University of London, my home institution, for granting me the period of research leave to undertake this position. While in Tokyo, many individuals shared their wisdom and their knowledge, and this book could not have been written without the healthy research climate that the Institute has been able to create. In particular, Akio Takahashi, Yasushi Oki, Masashi Haneda, Jin Sato and Yasuhiro Matsuda; Norihasa Baba and Michael Schiltz; and many others made me feel at home and provided great aid and encouragement. Ichiro Kakizaki, Ryuji Okudaira, Kei Nemoto and other colleagues in Japan at other universities were also a great resource on railways or Burma. Gratitude is also owed to my former dean and colleague Gurharpal Singh for allowing me the two years of leave from SOAS to undertake the Tokyo appointment.

Research for this book was mainly carried out in archives in the United Kingdom and Japan, with substantial fieldwork in Thailand and Burma (Myanmar) during this period. In the United Kingdom, the Imperial War Museum, the King's College London Liddell Hart Collection, the National Archives (formerly the Public Record Office), the SOAS special collections and the Asia and Africa Reading Room in the British Library were regular haunts. In Japan, the National Diet Library, the University of Tokyo Library and the Institute for Advanced Studies of Asia Library were also very useful for materials while I was writing. Additional visits were made to the Royal Engineers Museum in Gillingham (UK), Cambridge University Library, the Cambridge South Asian Archive, the Tokyo University of Foreign Studies Library and the University of Pennsylvania Library's Special Collections. I owe special thanks to Richard Duckett, who, during his research on the SOE in Burma in 2010, found several documents in the archives on the Burma Railways during the war, which he generously shared with me, and these provided important leads into some of the other material I have used in researching this book. I also benefited from the work of other scholars who have also researched different aspects of the First Burma Campaign, including Michael Leigh (SOAS) and Philip Woods (NYU London), and published recent monographs on it. Victor Lieberman (Michigan), Christopher Gerteis (SOAS) and Richard Reid (SOAS) all read and commented on drafts of the introductory chapter. Daud Ali and Ramya Sreenivasan (both at Pennsylvania) and Peter Wilson (Oxford) hosted seminars at their respective universities, where I presented aspects of the research for the present book.

Family oral histories and unpublished family archives have also been invaluable. Frank Biddulph, Brigadier Francis John Biddulph's grandson, showed immense generosity in immediately responding to a shot in the dark I took by sending out a letter on the basis of having run across his name by chance during the writing of the initial manuscript. He and his family opened their home to me and allowed me to go through the colonel's private papers kept in his old suitcase and to read them on the colonel's very own desk, a rare privilege. The suitcase was a treasure trove and included, among many documents of the 1942 Campaign not found elsewhere, the colonel's personal notepad wherein he sketched out his notes concerning the events unfolding around him. I have also benefited over the years from conversations with colleagues and students, future colleagues and, amongst the latter, Yi Li, the late Bianca Son, Ronnie McCrum, Thomas Richard Bruce, Thanyarat Aphiwong, Maung Bo Bo, Tassapa Umavijani and Kazunori Hashimoto have been inspirations in the energy and excitement they have taken in their research. Above all, gratitude is owed to my colleague and wife, Atsuko Naono, for her patience and her commenting on endless drafts.

Abbreviations

AA	Anti-Aircraft
ABRO	Army in Burma, Reserve of Officers
AFS	Auxiliary Fire Service
ARP	Air Raid Precautions
AVG	American Volunteer Group
BAF	Burma Auxiliary Force
BESA	British Engineering Standards Association
BFP	Biddulph Family Papers
BMP	Burma Military Police
BOR	British, Other Ranks
BR	Burma Railways
CBI	China-Burma-India Theatre of Operations
CLM	Chinese Liaison Mission
CLO	Chinese Liaison Officer
CRC	Chief Railway Commissioner
EinC	Engineer-in-Chief
EIR	East Indian Railway
ESO	Embarkation Staff Officer
FBC	First Burma Campaign
GHQ	General Headquarters
GOB	Government of Burma
GOC	General Officer Commanding

GOI	Government of India
HQ	Headquarters
ICS	Indian Civil Service
IE	(Royal) Indian Engineers
IFC	Irrawaddy Flotilla Company
IJAAF	Imperial Japanese Army Air Force
IMS	Indian Medical Service
IOR	India Office Records
IR	Indian Railways
IRS	Indian Railway Standards Committee
IWM	Imperial War Museum
IWT	Inland Water Transport
LHCMA	Liddell Hart Centre for Military Archives
LMSR	London, Midland and Scottish Railway
LofC	Lines of Communication
MCO	Movement Control Officer
MFO	Military Forwarding Officer
NWR	North Western Railway
OC	Officer Commanding
POW	Prisoner of War
PRO	Public Records Office
Q	Quartermaster
QMG	Quartermaster General
RA	Royal Artillery
RAF	Royal Air Force

RCC	Railway Construction Company
RCMG	Railway Construction & Maintenance Group
RMC	Railway Maintenance Company
ROC	Railway Operating Company
RTO	Railway Transport Officer
RWC	Railway Workshop Company
TSC	Transportation Stores Company
TTC	Transportation Training Centre
US	United States
WO	War Office
YBR	Yunnan-Burma Railway

Map 0.1 Burma Railways circa 1941.

Introduction

The First Burma Campaign (hereafter FBC) was one of history's greatest fighting retreats, from Tenasserim in the south up to India in the north. This story has been told many times from the perspective of commanders and soldiers, although there is growing literature on the roles played by colonial authorities and civilian populations.[1] A neglected part of this history was the success of railway operations persevering in the face of war up to the near completion of the Japanese conquest in May 1942. Although the Burma Railways (BR) played major roles during the campaign in both supporting the military defence of the colony and aiding the evacuation of civilians to the north from whence they could trek overland to India, the operation of the railway itself is almost never referenced and never discussed at length. This is the case, despite the enormous significance of the lessons of the campaign and the surviving BR staff on railway operations in India for the remainder of the war. They were influential, because when all other institutions had collapsed in Burma and everyone else had been withdrawn or were in retreat, the BR was just about the only thing that continued to work in what remained in the colony, in operations of rapidly diminishing scale. The subject of the book is how a specialist in running railways in war conditions per se, despatched from Britain, was able to get the railway running in the face of the enemy and why obstacles were placed in his path by the military establishment and railway managers in India and Burma.

One of the main obstacles to the effective mobilization of the railways in support of the war effort was the nature of interwar colonialism; however, the isolationism that reflected the impact of imperial and local colonial austerity measures in the post-1929 era, the impact of indigenous nationalism and political concessions made to it from the 1920s and concerns from the late 1930s in some Southeast Asian colonies over the Second Sino-Japanese War (1937–45) being fought in their neighbourhood would all contribute in their own way to the use of the railways in Burma's defence. The peculiar mentality of this period of colonial rule in Burma stood in great contrast to that of earlier decades of British rule there. Burma spent the decade up to the outbreak of the war to some degree hermetically insulated from many aspects of imperial affairs and those of neighbouring states. Heightened colonial isolation of the interwar years also played itself out in railway policy. The BR had periodically entertained the notion of cross-border linkages with China, Thailand and even through Assam to India, but these were abandoned. Burma's neighbours were certainly open to the prospect, with hopes that one could enter a railway carriage in Rangoon and then step out in Singapore, Bangkok or even Saigon. The Siam Railways and Federated Malay States Railways held annual conferences in the pre-war period in 1926 and recommended that the BR send a representative as well so that when an overland connection was made, the rolling stock would be of the right gauge to make interchange. Further, preliminary railway surveys to an overland connection between Burma and Malaya had already been conducted right up to the frontiers from both directions a quarter of a century earlier.[2] When the China–Burma Railway was finally undertaken (to complement the Burma Road), it was only after the colonial administration had fought tooth and nail to prevent it. From the 1920s, Burma sought to follow its own path separate from that of India.

Nationalism and Burmanization also played a major role in motivating the resistance of white railwaymen to militarization. The First World War had awoken a stronger sense of nationalism in Southeast Asia and elsewhere, recognition in many quarters of the validity of aspirations of national self-determination and the spread of new ideological tools to politicize, mobilize, and organize disenfranchised peoples everywhere. The threat that new anti-colonial nationalist movements posed to colonial and hence imperial security was coupled with more limited financial resources available to the empire, as well as the economic benefits of European rule to workers and cultivators, with the World Trade Depression which commenced in 1929. In interwar Burma, isolation and separation often overshadowed cosmopolitanism and international or imperial connections, while colonial administrators tried to keep populations divided and more strictly controlled their interaction and their exposure to new ideas.

Burmese had also developed their own imaginaries of what colonial Burma was, what precolonial Burma had been and what an independent Burma could be. These imaginaries formed counters to what Stephen L. Keck, drawing upon Arjun Appadurai's notion of scapes, has called the Burmascape produced by

British writers, many of them Burmaphiles, in the period between 1895 and roughly 1918. It was through the Burmascape that romanticized perceptions of Burma informed imperial and colonial decision making that then had a material impact on Burma and the Burmese through policy.[3] By the 1920s, however, Burmese writers, intellectuals and the small new middle class had produced their own Burmascape that informed how they and other Burmese would react to colonial rule and their own Burmascape was increasingly hostile to various things associated with foreign rule and domination. This included Indian immigration and the legal and economic advantages Indians and other foreigners were believed to have had that encouraged intermarriage by Burmese women with non-Burmese men and the feared extinguishing of the 'Burmese race' by being bred out of existence. Burmese nationalists thus demanded that Burma not only cut the formal 'webs of empire'[4] but also assert stronger control over its informal ties. In the case of Burma's relationship with British India, Burmese nationalists wanted not only constitutional separation and an end to formal links but also a snipping of informal links with India by reining in Indian immigration and the legal and economic advantages Indians and other foreigners had. Colonial authors writing in the 1920s and 1930s thus saw a different Burma than did their pre-First World War predecessors. Among these was J. S. Furnivall whose *Colony Policy and Practice* saw colonial populations as diverse, but they did not mingle and their religions and cultures coexisted, but did not mix.[5]

The circulation of military knowledge between Britain and India and Burma in 1941–2 had also lapsed for much of the interwar period. The Japanese offensive against Southeast Asia from late 1941 would change this, but certainly since the World Trade Depression from 1929, there had been no diapasonal circulation of military or related knowledge, approaches and technology because of colonial insularity within the empire. The transportation reformers examined in this book provoked resistance because their technique threatened to transgress territory onto which others had already staked claims and upon which their livelihoods depended, in particular in the 1930s economic and political climate in colonies like India and Burma. Gas masks and chemical agents were new elements in war and competed with no established technical or economic group for a place on the battlefield, however much they may have offended the sensibilities of combatants. Adopting jungle equipment and tactics may have led to groans and fevers, but only among the soldiers under orders; no civilians would have had cause for complaint. Nor did improved medical treatments threaten the profits of anyone other than undertakers. By contrast, changes to transport, especially the railways, threatened the economy of colonies and the social and employment positions of European technical and managerial elites in places like India and Burma. Behind Burma's 'walls' were colonial railwaymen who depended for their status on resisting imperial influence and control.

Colonial insularity also influenced local military resistance by the Indian and Burmese General Headquarters (GHQs) to changes in military transport. There is a growing literature that has sketched out the range of differences between the British and Indian Armies, but some devices were available that kept Indian commanders abreast with doctrinal and tactical changes in Europe, one of these being the Staff College at Quetta where even Gen. B. L. Montgomery became Chief Instructor in 1934.[6] Ashley Jackson has argued in *Distant Drums: The Role of Colonies in British Imperial Warfare* (2010), for example, that the British Empire was generally very well arranged for the mobilization of imperial military resources and cooperation. In his view, imperial networking kept elites scattered throughout the empire familiarized with each other, institutions provided an organizational basis for mobilization and communication, and an extensive transportation and communications infrastructure provided the physical means for doing so.[7] Overseas service, he suggests, became a common experience for the empire's service people in a military sphere of imperial life that could be characterized as a 'world of incredible movement'.[8]

Nevertheless, GHQ India was additionally slow to change because of the double perception of colonial insularity and local peculiarity. As a result, it did not accurately assess the potential operational context that India might be placed in should war with Japan break out. Although military Transportation had wartime experience in Europe, military authorities in India and Burma viewed it in 1941 as inapplicable to the colonial setting. In their eyes, the performance of military Transportation in warfare in the major industrialized societies of the northern hemisphere provided no lessons that were applicable to the greater Indian environment, ignoring for the moment India's growing commitments to the Middle East. Only local experience mattered and the local military establishments and the colonial transport agencies already had sufficient amounts of that. To a certain degree, such colonial authorities were indeed wrong. As the FBC would demonstrate, in terms of wartime transport services, Burma shared many of the same problems with larger countries in Europe in the Second World War. Its roughly 360 locomotives, 949 passenger coaches and 9,805 freight cars underwent grinding attrition inflicted by a Japanese bombing campaign.[9] Burma also had to balance out civilian and military demands on the railway, including the evacuation of population centres close to the front lines or especially exposed to bombing. The BR network was the site of much of the actual fighting. There was a constant fear of sabotage by the general population, looting of trains, signal and communications failures and train collisions. Most of these experiences were not unique and were typical of any railway system in the war that was under attack.

The BR also had unique colonial structural features that influenced its performance in wartime. Like the transport systems of other colonies, Burma's railway was exclusively oriented towards an extractive, colonial economy rather than a more diversified, industrialized economy and this had important implications for the organization of the railway and the arrangement of its technical aspect. The BR, for example, would remain bottom-, southern- or, perhaps better put, coastal-heavy, the bulk of its capacity oriented around the port of Rangoon. Nevertheless, more subterranean factors had a greater impact on railway operations in the FBC. Most significant was the resistance of local railwaymen to outside interference with railway matters. These colonial railwaymen did so mainly in reaction to developments that had occurred because of changes in colonial policies introduced in the previous two decades. As was the case in many colonies, racial hierarchies from the 1920s had begun to see slow but certain amendment. Burma was in the midst of its own policy of gradual indigenous empowerment, in which power was being transferred to the Burmese, and Burmanization in the civil administration, which meant the employment of more Burmese in positions formerly and exclusively held by outsiders. The ranks of the colonial railway officers would no longer be filled exclusively (and soon, not at all) by white men brought in from Europe or from elsewhere in the empire. White railwaymen responded to the gradual erosion of racial privilege that followed by emphasizing ever more strongly their membership in an occupational status group.[10]

A heavier reliance on membership within the body of BR officers meant (1) closing ranks to outsiders and (2) seeking insulation almost exclusively in the possession of peculiar authority that came with local experience in and knowledge of running the railways in an extra-European society. While open to implementing external imperial directives and externally modelled reforms themselves, as they attempted to do in the late 1930s and into the early years of the war, the BR administration was allergic to any direct interference by outsiders in colonial railway operations and resisted such efforts fiercely. Worryingly, the war and the arrival of external imperial authorities threatened to undermine their status. Unable and unwilling to adapt to a new way of doing things regarding the railways, BR heads resisted doggedly militarization by anyone outside of their own administration. This hurt railway operations perhaps more than anything else and very nearly brought railway support for the Burma Army and imperial forces in Burma to an end months before they actually did so. The military could only resolve the impasse by tearing the colonial railways' administration apart and releasing the railway and the railwaymen for reorganization into what became a military transport organization. However, like other railway companies it was largely populated by local, civil railwaymen who benefitted from years of experience within the colony and with its railway. The new Director of Transportation found it necessary in response to the exigencies of war to allow local railway managers to operate sections of the railways based on first-hand knowledge derived from local experience with local conditions. This book views this change as key to the continued functioning of the BR until the last days of Burma's defence in the FBC in 1942.

The present book is focused on the role of the main figures who delivered change in the face of colonial resistance for two main reasons. First, the book reconstructs a longer, more difficult and less predictable process than has usually been identified in the literature on either military adaptation or the imperial circulation of knowledge. The adaptation that occurred in India and Burma resulted from military–imperial contests that represented hard-fought political battles inside military institutions, colonial government departments and civilian transport agencies because of the serious consequences involved. During war, soldiers and/or civilians on either or both sides of the front lines would die because of changes made in technology and tactics, and this was a matter second only in importance to the military careers at stake. Second, because of the highly technical nature of Transportation Technique and its failure to excite the average imagination, the success of its adaptation depended more heavily than many other types of change on the role of technical specialists committed to promoting change to adaptation. These men and women were scientists in laboratories, tank or infantry instructors in training camps, field officers or filled any number of other roles where they could gauge the efficacy of manual and instructional learning and procedures with laboratory or real-time observation.

To deliver rather than merely recommend change, however, such technical specialists necessarily had to have a particular kind of personality, one that, once convinced they had a better way of doing things, actively sought to make changes according to solutions they had learned or devised. Those that took the added step were something else. Like junior officers in the field who took the initiative on the basis of their own appraisal of conditions, they politicked on behalf of their beliefs. In terms of Transportation Technique as discussed in the present book, two such individuals, both Royal Engineers, stand out for their roles in Britain, India, Burma and even Iraq in 1941–2. One was Lionel Manton and the other was one of Manton's instructors at Longmoor, the aforementioned Biddulph. In the case of the military railways, improved operating Technique had begun at Longmoor Railway Training Institute in Britain before the war. Biddulph carried it afterwards to India in 1941 and while being put in place in Iraq and Iran (then Persia), it was propelled with full force into Burma to respond to the railway failures there in the midst of a Japanese invasion.

Andrew Goss, in his study of the introduction of Western botany and Linnean taxonomy into the Dutch East Indies, refers to an earlier generation of such personalities as messengers, calling them 'apostles of the Enlightenment'. Biddulph several times referred to himself as a prophet.[11] In common parlance and in private ruminations there is not a great deal of difference. Technically, however, an apostle merely relays a message, whereas a prophet has been raised up by God and sends out apostles with his message. The present book favours the view of Biddulph as a committed reformer, even a prophet. He not only carried his particular message of improvement, but also actively pursued adaptation and used operational situations to test and improve upon the military Transportation Technique that he believed was the solution to not only India's defence against an expected Japanese invasion, but also would be a deciding factor in the re-conquest of Burma.

This was not an easy task and was not guaranteed success. Often, such reformers did not have the rank to provoke change on their own or the political skills to win, however high-ranking generals, on the merits of their case alone. Such reformers faced entrenched, senior commanders who saw no need for change and they were just as often put into their positions by a system that valued the existing way of doing things. Men like Biddulph thus experienced significant difficulties in pushing reforms into adaptation. As mentioned, in both India and Burma, both the army and colonial authorities fiercely resisted the introduction of military Transportation as developed at Longmoor. They believed it was not something that would work in India, nor would it work in Burma. Only by breaking up the civilian railway administration, putting the railways in Burma firmly under his control and militarizing cooperative elements of the BR men did Biddulph demonstrate what a modern military Transportation organization could do. The railways rose up again, kept the army going and got civilians out and stayed in action two and a half months longer than the civilian railway administration had claimed was possible. This was what military railway training at Longmoor had lacked, a test in a non-European, colonial setting that led to tweaking and adaptation that surviving BR men fed back into instruction in India. Military railway training had thus undergone a transition from a metropolitan technique to an imperial one. It is also arguable that, by contrast to other elements of the defence, it would be difficult to view the performance of the railways as a failure.

This was, however, only half the story. The local military and railway authorities who resisted Biddulph were partially correct. Transportation Technique as taught at Longmoor was not fully prepared for the extra-European environment. Before coming to Burma, Biddulph does not seem to have fully considered what the difficulties in applying it locally outside of Europe might actually be. In this sense, the campaign was a revelatory experience. It allowed Biddulph to learn a colonial railway, what made it tick, as he watched it collapse around him. The FBC was a virtual vivisection of the BR, exposing the living parts of the colonial railway and with it the colonial environment, the impact of colonial racial policies, exclusionary politics, economic disparities and limited investment in social welfare, to metropolitan scrutiny, through Biddulph's eyes, before each part withered and died. Only slow, grinding destruction inflicted by the Japanese permitted this and only this experience could impress upon Biddulph that Transportation Technique had to be adjusted.

An especially significant defeat has the potential to provoke significant changes to an army's tactics and technologies because of the trauma they inflict and the investigations, innovations, reforms and adaptations that follow.[12] More than anything else, defeat in the FBC was certainly a necessary part of gaining acceptance to reform and the adaptation that followed. Recognizing this, Biddulph played a key role in representing to Indian military authorities that, rather than experience success, the railways failed or would have failed had it not been for the continuing military reverses of the FBC. This seeming contradiction will be discussed in Chapter 8. Biddulph found making a clear case of BR failure, however, extremely difficult and hard-fought due to the firestorm of blame and acrimony immediately following the retreat as military commanders and colonial officials wrote angry letters, self-serving reports and, later, memoirs that explained their actions over and over again. The FBC in some ways was a repeat performance of the same retreat experienced at Dunkirk in 1940 by at least some of the same commanders. While transforming Dunkirk from a military defeat to a miracle was a feat of wartime propaganda, doing the same for Burma was seemingly impossible. Calm, rational assessments were rare because the chaos of the campaign, especially in the last few months, made it difficult for anyone at the time to have had a sufficiently broad view to gauge what factors really made the difference. Biddulph succeeded in the end, but suffered personally and for a short time, professionally, in the process.

The literature already recognizes that the military defeat of the FBC served as a catalytic event in promoting a range of military reforms in India that contributed to the military victories in northeast India and Burma in 1944 and 1945, respectively. But the literature has mainly focused on those that related to the infantry, strategy or military supply per se. Daniel P. Marston, for example, has shown how infantry reforms including those related to jungle warfare were channelled into new training memoranda and military training pamphlets being published even before the FBC was over.[13] Tim Moreman has shed light on changes in doctrine.[14] In Graham Dunlop's important study of military logistics and related aspects of the Burma Campaign(s) generally in the Second World War, he looked at the FBC to establish the broader organizational and systems aspects of logistical problems at the strategic, operational, and tactical levels. This was intended as a prelude to examining how these problems were resolved in supporting the army over the course of the remainder of the war. However, Dunlop devotes little space to the operational procedures of elements of transport, such as the railways.[15]

The roles of Manton, Biddulph and indeed the BR men in this campaign, however, are largely but not completely unheralded in the literature. One factor obscuring the role of these men and indeed of the railways in this campaign was a missing cache of sources whose absence obscured the role of the Transportation Directorate, which coordinated the management of all transport in the colony during the FBC. These sources were only recently located by the author in the possession of Frank Biddulph, the grandson of Francis John Biddulph, the Director of Transportation in Burma during the FBC. This material was kept in a suitcase and included the elder Biddulph's personal correspondence, diaries, reports and so on. By contrast to the silence of Biddulph in the India Office Records, the Imperial War Museum, the Royal Engineer museums, the Liddell-Hart Collection at Kings and a range of other archives, this suitcase, unopened in the years since his death, yielded a treasure. Moreover, among Biddulph's own papers were a collection of other unpublished reports and letters from participants in the transport operations of the campaign. There are reports on running the loco sheds, the railway yards and the railway's situation in Rangoon, and a range of other sources apparently sent to Biddulph during the campaign. Within this collection are included a range of voices from Transportation officers and railwaymen whose reports on the performance of the railway in the campaign remained in Biddulph's suitcase. The suitcase is in other words an essential archive for understanding the Transportation Directorate's and the railways' roles in the campaign. No previous historian of the Burma Campaigns has accessed this aging leather archive. Most importantly, these sources help us to bridge the information gap and clearly show the details of how the military's relationship with the railway in the FBC served as such a catalyst for reforms in the Indian military Transportation establishment from 1942.

As mentioned, the ability to keep the railways in operation despite Japanese bombs and bullets permitted Biddulph to test out and further develop state-of-the-art military Transportation Technique as taught at Longmoor Railway Training Institute in a wartime setting. This experience brewed metropolitan military Transportation Technique together with actual wartime experience in a colonial setting that would allow it to overcome the resistance of the Indian Army Command and become hugely important in the formation of India's modern military Transportation establishment. The crisis produced by defeat in Burma aided evolution, Biddulph using its flames to get his message across when he returned to India in May 1942. Veteran BR men and Indian Engineers (IE) also brought their experience in military Transportation in a war zone and disseminated it both as instructors at the level of staff training institutes and as the cadre of new military railway formations in India.

In terms of timing, the quickest change regarding transportation that took place in India after the FBC was the development of the railways in Assam. This reaction was not difficult to explain. Historically, India had viewed her main area of military vulnerability as the northwest frontier and thus her logistical infrastructure was concentrated there and not in the northeast and this had been one reason why India had been called upon to be responsible for the transportation needs of PaiForce.[16] However, the development of the railways in Assam overlapped with and was connected to changes in the railway force. Training centres that had very slow beginnings in 1941 and which the Quartermaster Branch in India unenthusiastically tolerated now mushroomed in size and the intensity of the training. Some BR men were put in charge and made instructors, other BR men were brought into army headquarters and still others were placed on the Assam LofC as the commanders of new military railway units. They put in place military Transportation training from Britain that they had learned while they were on the footplate or in the yard in Burma while under Japanese attack. They were the carriers of metropolitan military Transportation Technique and were the vanguard of changes in India's military railway operating. They would be at the centre of logistical support for the building up of the Indian Army, its defeat of the Japanese in northeast India in 1944 and its successful counter-invasion of Burma in 1944–5, and would pioneer the modern Indian Army's transportation elements.

The chapters are arranged semi-chronologically. This was necessary because as each stage of the FBC changed the quality and scale of the pressures placed on the railway, different features, including both strengths and weaknesses, were brought to light as the campaign progressed. The subject of Chapter 1 is the development of military railway Transportation, what was sometimes referred to as military Transportation Technique, by the 1930s as a metropolitan military approach and the difficulties several military reformers faced in encouraging its adaptation in several venues in the empire. One of the biggest obstacles was that commanders for the position of Quartermaster General (QMG) in India, chosen based on their administrative experience, training, and skills that were admittedly essential, nevertheless lacked familiarity with the principles or importance of military Transportation. This meant that the situation of the movement of military Transportation through the stages of innovation and adaptation in a colonial context, in this case India and Burma, was very different from the cases of chemical warfare in the First World War and the development of military medicine as examined in several very highly regarded studies. Albert Palazzo and J. P. Harris, for example, have shown that British generals were very open to the implementation of scientific responses to chemical warfare and the introduction of tanks.[17] Mark Harrison has shown that military medicine's trajectory of development during the Second World War owed much to the 'command culture' resulting from the rise to leadership of high-ranking military officers already convinced and committed to it.[18] The situation of military Transportation Technique, however, could not have been more different. Outside of the Royal Engineers and sometimes even within this august corps, there were few who understood the details or value of military Transportation and few among the generals in India who would champion its cause. Often Biddulph and other reformers found themselves fighting these generals as much as they did anyone else. Chapter 1 reveals the politics of this fight whose gradual resolution will be the focus of the chapters that follow.

Chapter 2 considers the question of why these railwaymen were so resistant to outside interference, why men who would evince so much group loyalty would be among each other's worst professional rivals and how tensions and bonding among them, seemingly contradictory passions, both became especially powerful on the eve of the Japanese invasion. It will be argued that an era of indigenization (or Burmanization as it was called in Burma) of colonial bureaucracies meant that many of the white, Indian and Anglo-Burman railway officers in Burma had to rely more heavily upon their local expertise in railway operations in the local colonial context for their command of their position, status and authority over railway operating. As the colonial political climate continued to erode a guaranteed claim to status based on race from the 1920s, a new threat emerged in the 1930s from Depression-era retrenchment efforts. Retrenchment made insecurities among railwaymen worse by threatening chances for professional advancement, and for particular individuals, these prospects seemed bleaker because of a new administrative scheme introduced just before the outbreak of the war.

Biddulph's arrival and his effort to reorganize BR operations according to metropolitan military Transportation Technique is the focus of Chapter 3. The introduction was slowed, however, because of an entrenched civilian railway administration that resisted change and claimed operational competence because of fears of outside interference in a changing colonial political context. This resistance persisted despite its inability to deal effectively with serious manpower shortages and demands for Forward Support, and as performance declined, tensions emerged between military and civilian users of the railways that could not be resolved. As Biddulph's efforts showed, it was by no means a given that innovation would be accepted by those in need, even when it was very clear that the current ways of doing things were failing. Change threatened to upset the dominance of local, colonial elites over the railways and it was the resulting tensions and resistance that delayed the introduction of military Transportation Technique until it was perhaps too late to save the situation.

Chapter 4 turns to the reasons why Biddulph was afforded an opportunity to fully introduce Longmoor Transportation Technique and Transportation organization to Burma, over two full months after the war with Japan had broken out and just as the Japanese were about to move into the Irrawaddy Valley. The different stages of militarization involved first mobilizing a group of BR men who had lost out in institutional competition in the late 1930s in support of the Transportation Directorate by commissioning them and putting the railways under their charge. Although the civilian railway administration was banished to upper Burma, the latter would also see railwaymen militarized. Militarization aided railway operations significantly and got them running again until the end of the campaign.

Chapters 5 to 7 broadly examine the performance of Burma's militarized railway and its continued survival while the defence of Burma faltered and then failed. Chapter 5 begins this discussion by explaining how the decentralization of authority in railway operating during the FBC proved crucial for continued railway survival. As in the pre-war period, the problems of running the railway during the FBC were not uniform across the network nor were they static in any one geographical area of railway operations. Railway operators responded to different demands that changed in quality and intensity as the campaign progressed with performance that was determined by the changing reserves of manpower, motive and fuel resources at its disposal, and these concentrations were heavily localized and ephemeral. They required local experience and local assessments to manage. This chapter examines how the FBC affected a group of local railway managers on a small portion of the railway network that nonetheless experienced some of the most dramatic challenges to railway operations. They had to do so largely without central direction in part because of communications breakdown and physical isolation from Rangoon from late February 1942. However, the decentralization of authority was a choice Biddulph consciously made for the railways once he had effectively deposed the Chief Railway Commissioner Chance. Biddulph wanted local, militarized groups of railwaymen to operate on their own regarding local conditions in pursuit of central directives.

Chapter 6 continues the discussion of the essential importance of relying upon local experience and knowledge, in this case, the BR's technical and linguistic aspects, to inform Transportation Technique. The legacy of colonial railway policies had a real impact on their ability not only to conform to standardized procedures but also to support military operations per se. Nevertheless, there were also things that local railway operators could use in operating trains in a wartime setting for which pre-war local railway operating did not prepare them.

Chapter 7 looks at how small groups of railwaymen maintained local railway operations throughout the last weeks of the FBC. The challenges and obstacles of this period were an omnibus of all the problems seen in the campaign for railway operations. The period and the continued performance of the railways on the Myitkyina Branch helped to assure that attention would be paid by New Delhi to the importance of the railways. The prolonged example of a railway apocalypse made it possible to sustain the 'failures of the railway as catalyst for change' approach Biddulph and Manton had adopted since February 1942 as a means of keeping up pressure for the militarization of the railways in India. The period also explains why, as will be discussed in Chapter 8, Biddulph and Manton were so committed to recruiting the BR men for military training at Deolali and their mobilization for military railway operating and construction companies in India afterwards.

Chapter 8 is interested in the impact of the BR men's and the Transportation Technique's shared experience in the FBC. It begins with a discussion of the criticism of the Transportation Directorate in Burma and why the commitment to a stronger military Transportation and Movement Control in India was one outcome of the FBC nevertheless. It is argued that this was the result of campaigning by the reformers and 'prophets' of military Transportation. Adaptation in the case of Transportation Technique had been a long and difficult process in part because of opposition rather than aid from the top of particular branches of the Indian Army. The BR men who wed military Transportation together with local colonial knowledge and experience in wartime put the Technique into practice in military railway operations. This fully transformed what had been a metropolitan Technique into an imperial one. The conclusion examines the key findings of the book and its implications for the history of military adaptation and the imperial circulation of knowledge. The chapter shows why the impact of imperial Transportation (and Movement Control) proved durable and developed into an all-purpose military method that would aid global, British military transport in deployments in later generations.

1

Metropolitan Transportation Technique from Britain to India and Iraq

With the South African War (1899–1902) had come the realization by the British Army that Britain's wars were now shifting from the asymmetrical warfare meted out against a host of indigenous societies in Asia and Africa to those against armies comparably armed and societies much more able to wage modern war. This also meant that railways, although sometimes merely convenient in the past, would be essential in future wars in areas of the world where river transport was not available. Military railway traffic control and operation thus emerged during this conflict as an important logistic tool.[1] Although immediately set aside after the war, it was soon picked up again. After further development and institutionalization in Britain, it evolved into military Transportation Technique, often referred to simply as Transportation. Although at first a metropolitan technology, this body of training was soon indirectly shared with the empire through joint imperial military campaigns by the late 1930s and Indian Army control of the railways in the Middle East from 1941. The carriers of this training, transportation officers from Britain, were not entirely welcome in India in 1941 nor were they in January 1942 in Burma.

One of the biggest obstacles to the adaptation of Transportation Technique in several venues in the empire was that commanders who held the position of QMG were chosen based on their administrative experience, training and skills that, while undeniably essential for many aspects of their role, nevertheless lacked familiarity with the principles or importance of military Transportation. There were few among the generals on the other side of the table who would champion its cause. Often Biddulph and other 'prophets' of reform found these generals to be their greatest adversaries. The present chapter focuses on the politics of this fight whose gradual resolution will be the focus of the chapters that follow.

Longmoor and military railway engineers

Indian military logistical development had not kept pace with the evolution of British military Transportation Technique as taught at Longmoor by the time of the outbreak of war. British Army manuals that Biddulph brought to Burma outlined procedures and his own limited hands-on experience with war conditions in Iraq informed Biddulph. Biddulph would be accompanied and followed by other officers who were also posted to Iraq just prior to the Burma Campaign, including an Indian Army officer, Col. James Newton Soden; Lt. Gen. William Slim, who would be flown in and made commander of newly formed BurCorps in March 1942; and Brig. T. O. Thompson, who would also be flown in from Iraq in January 1942 to replace Brig. E. C. Lang as the Deputy Director of Medical Services, Burma.[2]

In terms of training for the military operation of a railway, no British colony had anything comparable to Longmoor. In Britain, the Directorate of Movements and Quartering under the QMG to the Forces was charged with two main tasks related to war. The first was to plan for the movement of the Field Force on mobilization and to collect information of possible theatres of war overseas, and the second was to impart technical knowledge so that railway troops could construct and run railways where the army needed them. In Britain, a special centre, the Railway Training Centre, Royal Engineers, was established to train military railway operating companies (ROCs). In 1905, the Bordon and Longmoor Military Railway was built to meet this need, although authorization came after the fact in 1906. In 1908 it was named the Woolmer Instructional Military Railway and in 1935 it was renamed the Longmoor Military Railway. Here was a permanent way, railway yards, railway workshops and the like to provide training to military engineers to operate the railways in the way the military needed. The Longmoor courses were two weeks in length, equally divided among military and technical training, and focused almost entirely on railway work. As one account described the training in 1935, they covered everything from heavy bridging to railway signalling.[3] Among the other important lessons taught at Longmoor were the combined exercises between other arms of the military and the railway troops. An important function of the centre, in other words, was directed at helping formations work with a military-run railway, providing space for them to 'rehearse entrainment and co-operation with railway troops in the special problems of transportation'.[4] In Britain, however, understandings of military railway engineering into the late 1930s were still shaped by the experiences of the British Army with working the railway network in Belgium and north-eastern France to keep the British Expeditionary Force and other imperial forces in supply on the Western Front in the First World War.[5]

One of the instructors at the centre was Biddulph, who began teaching there in 1934. In 1937, Biddulph would publish an article on signalling instruction at Longmoor in the technical journal, the *Railway Gazette*, which would summarize the Signal School's and hence Longmoor's main tasks. These were to train the railways' tradesmen of the Royal Engineers – the blockmen (the military equivalent of signalmen), firemen, engine drivers, shunters and brakemen – 'in rules and regulations for the safe working of a railway'. It would also teach the application of railway operating principles to officers and traffic operators (the equivalent of yardmasters, stationmasters and train controllers) and the elements of mechanical interlocking to signal fitters.[6]

There was a No. 0 gauge model railway arranged like a giant horseshoe in a large hall for the Signal School's pedagogical purposes. Students operated it with a diagrammatic electric switchboard with the coloured lines on the board corresponding to coloured lines along the model railway's vertical side panels. Stations on the layout were constructed so that they provided the students examples of the various kinds of railway station layouts they might encounter in the field. These station layouts (named Windsor, Slough, Langley, Oakhanger, Whitehill and Woolmer) were arranged in order of increasing complexity as one moved around the layout. As instruction progressed at Longmoor, Biddulph found that he was also able to use the model effectively to instruct students on how to marshal trains, shunt and undertake more advanced operating work. Controllers also learned their duties by running the model trains according to a graphic timetable. Accidents that blocked the line on the model railway were also staged 'to encourage initiative and resource in dealing with situations … liable to arise' in a war situation.[7]

Biddulph came to Longmoor at a time when there was demand, after a long period of neglect and more circumspect contingency planning, for the British Army to be both economical and flexible in its application, for possible overseas deployment not just to continental Europe, but also in other theatres. Britain's imperial defence arrangements, what James Hevia has called 'the British imperial security regime', directed by the War Office (WO) in London, had to prepare for two different kinds of warfare, 'civilized' and 'irregular', by the beginning of the twentieth century.[8] The first threat was dealt with in the First World War with the defeat of Germany. After the First World War, the British Army adopted the ten-year rule, whereby defence planning was geared to the principle that there would not be another war requiring a British expeditionary force in Europe for another ten years. Instead, Britain should only prepare for extra-European deployment, against enemies who could not have been expected to compete with a modern European Army. This period of 'severe economic restriction' lasted long enough to become a 'habit of mind'.[9] As the 1920s progressed and the World Trade Depression followed the New York Stock Exchange collapse in 1929, the ten-year rule was renewed annually, a position that began to change only after Japan withdrew from the League of Nations in 1933, the rise of Adolph Hitler in Germany in 1933 and the Italian war in Ethiopia. The army put in place new priorities from December 1937 that emphasized imperial commitments, including preparation for the expeditionary force to be used outside of Europe, such as in Egypt, coupled with a limited liability policy.[10]

An effort was made through the production of such manuals to provide guides to British officers on everything they needed to know to manage different responsibilities and to ensure uniformity in organization and practice throughout each military service. There had been such manuals on how to manage a military railway in the First World War, and the interwar period saw new manuals emerge for Transportation and Movement Control as well and, as we shall see, for Movement Control in war conditions in an overseas context. These manuals were necessary because of the multiplicity of railway companies (numbering 120 at the time) in Britain until the 1920s, when the Grouping Act (the Railways Act of 1921, put in force in 1923) created four large railways, the 'Big Four', the GWR, the LMSR, the London and North Western Railway, and the Southern Railway.[11] There was general standardization across all the lines in railway operations and engineering because of the work of the Railway Clearing House (RCH), which had originally been established in the mid-nineteenth century to manage fare sharing when interworking began among the railways in Britain. Although this remained its main purpose the RCH also became a vehicle for effecting standardization through the issue of the *RCH Standard Book* and the *RCH Standard Regulations* (the first edition was issued in 1872) that established regulations agreed across the industry, creating a de facto British way of running the railways. On the basis of the RCH standards, each of the Big Four had issued their own railway rule books, albeit all in general agreement with acceptable nuance. This situation would not change until the Big Four were nationalized in 1948 according to the Transport Act of 1947 and a singular *British Railways Rule Book* was published in 1950. Thus, even leaving aside the special military demands on railway operation and engineering, manuals and rule books were needed prior to 1950 if only to provide a single set of rules.

Much of what was taught at Longmoor regarding military railway operation was codified in the *Military Railways Rule Book* (1938) (which drew upon, wherever necessary, the *RCH Standard Book* and the *RCH Standard Regulations*), which was one of the core texts of the military Transportation Canon. It embodied Transportation Technique as far as railway operations went, providing a standard set of rules and regulations military railwaymen should follow regardless of their experiences with other, civilian railways.[12] There was also a larger series of post-canonical manuals for teaching new military railway staff or as a reference for those who had already been trained. These manuals were produced by the WO and published over the course of the early years after the outbreak of the war in Europe. The roughly half-dozen volumes (the issue of supplements makes precise enumeration difficult) specifically directed at military railway engineering are volumes dedicated to railway engineering in general, surveying, railway bridging, railway operation and all the other general principles and instructions that were considered essential for running a railway for military purpose. With charts and diagrams, detailed instructions and carefully defined procedures, the tone of the discussion was directed at broad application, identifying the core methods of railway construction and operation as applicable anywhere the army might go. The aforementioned *Military Railways Rule Book* provided the basis for the updated *Notes on Military Railway Engineering Part IV: Operating* (1942), which best represented the state of Transportation Technique regarding railway operating at the time Biddulph would arrive in Burma.

What the military railway engineering volumes and the *Military Railways Rule Book* and even the training exercises at Longmoor lacked, however, was recent or direct experience with these activities in extra-European settings among those in Britain producing the manuals. The literature on other mechanisms of imperial control regimes, such as intelligence gathering, has demonstrated the interrelationship between colonial agents and British government departments, the latter providing the detailed local data to government agencies that then funnelled this information into intelligence reports that then re-circulated out of the centre. It is unclear, however, what, if any, means should have been undertaken by the QMG regarding preparing for the possible provision for railway (or other transport) services during much of the interwar period. Railway operation was mainly viewed through a British lens. As the First World War showed, when the British military ran many narrow gauge lines to support the British Expeditionary Force in the trenches, there was no major knowledge gap regarding railway construction and operation per se. Hence, Longmoor instruction was geared towards operating the railways in a war in a British or continental setting, with some accommodation of specific overseas experience in living memory, such as the operations in Mesopotamia (Iraq) in the First World War.

With the new demands placed on the British Army in the mid-1930s, however, there was indeed now a knowledge gap and there was no centralized system in place to collect local and detailed data on the actual experiences (as opposed to technical procedures) of railway operations overseas. Instead, there were three main sources available to gain insight. The first two sources consisted of the annual railway reports that provided mainly statistical data, very general and brief qualitative data, and the odd interesting event that the compiler considered sufficiently worthy to include. The second source consisted of the more in-depth but still very abstract publications on particular matters issued by the major colonies, such as India, and the dominions whose railways were of a scale sufficient to legitimate the costs of their research and costs of production.[13] Such materials, however, did not aid much in instructing railway engineers in what the situation might be like in smaller and poorer British colonies or foreign lands. The third source was the vast array of private sector publications directed at the huge international community of railway operators, investors and equipment manufacturers, as well as a large community of train-watchers, train-spotters and *tettsu-chans* (in Japan). These publications included such monthly magazines as the *Railway Gazette*, *Railway Progress*, *Locomotive Magazine*, *Railway Magazine* and a host of others, and the WO's military railway engineering manual compilers turned to these magazines for help. As a result, these manuals' examples of extra-European experiences were mainly drawn from references to operations in Mesopotamia in the First World War reprinted from issues of the *Railway Gazette*.[14]

The outbreak of that war in 1939 would reveal how much overall Allied strategic mobility depended on sea movement, in addition to the continuing dependence on rail and road movements. Although familiarity with knowledge of shipping movement was encouraged, no attention was paid in Movement Control training to air movement and very little to road movement. It was the case that the military Transportation organization now involved more than just the railway. Good military Transportation officers now needed more specialized and broader knowledge and there had to be a reconfiguration of the division of labour among transportation staff. Military-trained staff officers needed to have an understanding of the working and needs of military formations. Technical officers needed to understand the operation of multiple means of transport and to be able to communicate in their respective operational languages. However, both had to have an awareness of the work of the other in order to effect smooth collaboration.[15]

Nevertheless, up to 1944, instruction at Longmoor would remain focused on the railways. As the training stressed, military railways were responsible for carrying troops and supplies along the lines of communication (LofC) from the Rear Maintenance Area to the railhead in the Forward Maintenance Area. Technically, as the instructors at Longmoor communicated to their instructees, a LofC could commence in Great Britain and stretch all the way over to the continent (or other overseas venue) and could include any kind of transport, 'but we normally look at it from the ports to the rail-heads'.[16] As a result, much of the hands-on experience focused on railway work and the loading and unloading of different kinds of railway wagons.[17] Railway troops also learned, of course, the reverse, how a military-run railway could entrain various kinds of military formations. How much training at Longmoor was directed at the operation of a railway in a front line context is not clear. Nevertheless, the kind of training undertaken provided a better coordination of railway service to meet military demands behind the lines. Instructors kept the students focused on the five key control points along the LofC relevant to transportation staff. These included (1) the docks, (2) the exchange sidings, (3) the depots, (4) the Base Marshalling Yards and (5) the regulating station points. Operating or transport control was responsible for traffic when it was 'on the move' between any two of these five points.[18]

Around the empire: From India, Iraq and Persia

What had been learned at Longmoor was not essentially imperial knowledge, although it was partly designed to prepare British formations for overseas expeditions and would arrange for one uniform set of operating procedures, published in *The Military Railways Rule Book*, that during wartime would 'obtain world wide, no matter how climatic conditions or administrative and operational circumstances varied'.[19] Although this set of procedures would become imperially circulated after the Second World War broke out, it would remain metropolitan in substance until it had changed through the process of circulation and adaptation to local contexts in the empire. The transmission of this metropolitan 'learning', or as Royal Engineers called it, 'the Transportation Technique', overseas relied on three routes. The first route was by the direct training of railway operating personnel at Longmoor, which would mainly have an impact on the Canadian military ROCs that came to Longmoor later in the war. Colonial Burma appears to have sent no one to Longmoor for training by 1942, so this route was not relevant to the FBC. The second route was again via the circulation of WO military manuals. Nevertheless, the first two of the half-dozen manuals on the principles of military railway engineering were only published in 1940 and do not appear to have gotten to Burma yet in time for the FBC. More to the point, the fourth volume, which would otherwise have been directly relevant to Burma's situation as it covered the actual operation of such a railway along military lines, was not issued until later in 1942, after the campaign in question was over. The third route was by the carriage of this learning overseas by British transportation and movement officers who might be sent overseas. Training at Longmoor and Derby would influence Burma by this route, via instructors who were despatched by the WO to aid transportation operations in India and Iraq. Among these, the two most important were Brig. Francis John Biddulph and Col. Lionel Manton.

Manton had originally been trained in Britain at the School of Military Engineering and at Longmoor after being commissioned in the Corps of Royal Engineers in 1906. In 1909, he joined the Military Works Service in India. As David Ronald and Mike Christensen explain, it was a practice in the Royal Engineers at the time to send their railway officers to India to gain practical experience. Manton stayed on and gained local Indian railway experience with the North Western Railway (NWR) for two years and later with the Bengal Railways, in both of the latter cases as an Assistant Traffic Superintendent.[20] With the outbreak of the First World War, he gained practical experience with the railways in continental Europe, the Mediterranean and North Africa for two decades. This included service in France in 1914 with the sappers and miners. He then became an Instructor in Transportation at the School of Military Administration in Britain after the war, followed this with a posting on the Rhine as the President of the Inter-Allied Railway Sub-Commission at Cologne and then spent time as the British member of the same commission and the Assistant QMG (Movements) for the Rhine Army. In 1930, he was promoted to Commanding Officer, Railway Training Camp (Railway Training Centre, Royal Engineers) – Longmoor Camp, Hampshire. Manton held this command until 1935, a period of four and a half years, including the first year of Biddulph's presence as an instructor. Army retrenchment in 1935 put Manton on half-pay and then into retirement. Manton even briefly took up a management position at Hodgkinsons in London, taking up what is described as a 'shabby semi' in Dulwich, but soon after found transport and engineering work in Egypt where he was made Assistant Director of Transportation for the British Army there in 1935 and as Chief Engineer in Malta, which has no railways, in 1936. In 1937, he was given what his family remembers as his 'dream job', being put in charge of the establishment of the LMSR School of Transport at Derby. Manton would remain the school's principal until 1939, when it was decided to close the school.[21]

Manton's chief concern was how to keep traffic moving under war conditions. In his view, railway operations had not caught up with the reality of the threat that aerial attacks posed. In 1938, for example, he criticized a transportation official for understating this threat, on the one hand, and 'putting the clock back' in proposed measures to offset potential disruption and damage. In Manton's view, suggestions that the railways needed to avoid concentrations of rolling stock and engines in running sheds were unworkable because of economic necessity. Manton was particularly worried about the demoralizing effect air raids had on railway staff and passive defence measures that could be used to reduce the lag in the time it took staff to get back to work.[22] When war did break out, during episodes of aerial bombing of Britain's railway yards in 1940, yard work could drop by as much as 70 per cent in a particular yard for the day, as measured by the total number of wagons shunted. The cumulative negative impact on effective work over the course of an attenuated period of aerial assault was staggering even where actual physical damage was not inflicted directly on the railways.[23]

Military railway training soon shifted out of Longmoor and back into Manton's hands. On Manton's recommendation, the WO took over the School from the LMSR 1940, as No. 2 Railway Training Centre RE (Longmoor becoming No. 1), with Manton as the commandant.[24] Manton's move coincided with the changing political situation in Europe. After Dunkirk, Movement Control training at No. 2 adapted to train Movement Control Officers for both home service and overseas. This would continue for another year under Manton's leadership until June 1941, when No. 2 was closed. The WO kept the associated line, the Melbourne Military Railway, in use as an extension of Longmoor for very nearly the duration of the war, its utility less important after the Normandy Landings in the summer of 1944.[25]

Manton's former subordinate at Longmoor was Francis John Biddulph. Biddulph's military career had begun in the First World War, being commissioned in the Royal Engineers as a 2nd lieutenant in late 1915, serving in France and Belgium on the Western Front from 1916 to 1917 and then in Italy for the remainder of the war. Clearly, this experience had an important impact on his confidence that military railways and other forms of transportation had a vital role to play in war and, at several points in his career, reinforced his points by drawing attention to the lessons asserted by important historical works on transportation work during the First World War. He was promoted to captain in 1925 and then major in 1934, becoming the following year, as mentioned, an instructor of military railway engineering at Longmoor, and chief instructor there in 1939. After becoming a lieutenant colonel and then a full colonel (acting), he was sent to the Middle East to advise on transportation matters.[26] Biddulph's position as an instructor meant that his public dissemination of military training information in the 1930s was limited to the *Railway Gazette* where he wrote highly technical pieces on aspects of instruction there, but he, like Manton, proved to be a very strong advocate of modern military Transportation Training or Technique in the years ahead. Opportunities to do so outside of Britain came in 1939.

With the outbreak of the war in Europe, there at first appeared to be no overwhelming demand for a full mobilization of Indian military resources for the war. Although the Indian Army considered the possibility that the QMG might need a transportation adviser because of its vulnerability on its Northwest Frontier, it took no immediate steps because it was not yet clear what Germany's strategy was. Biddulph, who was on tour in India when war broke out, urged that a transportation organization be raised in India as any study of military Transportation history would show how important the coordination of transport by the military had been on the outcome of military operations. Although he had recommended that at least the training of railway operating units be initiated, these suggestions made little headway and he was recalled to Europe in 1940 to aid with the transportation effort there.[27] It took nearly two years after the outbreak of the war for India to raise a military ROC or to form a Transportation Directorate.

The brief era of Indian Engineer ROCs had come and gone in a three-decade period from the beginning of the twentieth century. During the Boxer Rebellion (the Third China War, 1900–1), the various countries making up the Peking Relief Force were to each operate a section of the railways in Beijing, but as there were no British railway troops, the Russians did this on the British section on their behalf. When British railway troops did arrive, the Russians were hesitant to relinquish control of the section. Thus, as E. W. C. Sandes explains, there was a clear need for organized and equipped railway units and India converted some engineers into two railway sapper and miner companies and trained them in railway construction and operation. The creation of various railway battalions during the Second World War was followed by reorganization in 1921 and the abolition of the 'railway battalion' as a unit. All railway sapper and miner units save for Nos. 25 & 26 companies disbanded and the latter two companies were joined with the 3rd Royal Bombay Sappers and Miners. The units alternated with each other, spending one year on railway construction and the other undergoing military training at Kirkee. The financial impact of the World Trade Depression invited military entrenchment in India, there was no longer a perceived need for the role of railways in overseas expeditions and the two railway companies were ordered disbanded. Hence, the two railway companies were disbanded, with No. 26 being disbanded in September and No. 25 being disbanded on 16 October 1931. Instead, there was now a Railway Reserve Regiment made up of NWR employees to deal with trans-frontier railway construction.[28] This minimal military railway establishment would clearly be insufficient when India was called upon to serve as a base for the Middle East and Iraq in 1941.

Biddulph's second opportunity to influence India came in 1941 when he was despatched to the Middle East on a Transportation liaison visit.[29] The WO sent a telegram to the General Headquarters (GHQ), India while Biddulph was there, recommending that he be allowed to visit, before he returned to Britain, so that he could discuss the various transport problems that were emerging.[30] At this time in India, military Transportation efforts were seemingly headless and 'nobody's child'. Mainly the Engineer-in-Chief (EinC) Branch now undertook the effort with some help from the QMG's Branch (in India, the EinC was not under the QMG).[31] On the same day as Biddulph's arrival at GHQ India, 10 May, a telegram arrived from the WO suggesting that Biddulph might be lent to undertake the work as an Assistant Director of Transportation, advising the GHQ India on how to organize transportation services and employ transportation units.[32] The QMG and the EinC took this as a firm offer and Biddulph was appointed the Assistant Director of Transportation, India and the request to do so was made by the GHQ on 13 May.[33]

By 19 May 1941, Biddulph began to organize a Transportation Directorate in the QMG's Branch, immediately putting in for the provision of two Deputy Assistant Directors of Transportations. Biddulph then headed to Calcutta to meet with the Commissioners of Ports and other river and docks authorities to organize a Docks Group and Inland Water Transport (IWT) Company. Biddulph was relieved because of the opportunity and he now felt he had to develop a transportation organization in India. 'The wheel', he wrote, 'had come full circle, for the Transportation Directorate that I pressed for at Army Headquarters when war broke out is now taking shape. I hope better late than never.'[34]

This relatively new transportation organization was already overstretched. As an empire within the empire, India had multi-stranded relationships within the Indian Ocean as Thomas R. Metcalf has made so vividly clear in *Imperial Connections: India in the Indian Ocean Arena, 1860–1920*.[35] Just as the Indian Army was mobilized in support of British operations across most theatres in the war, India was also tasked in 1940–1 with logistical support in the northwest, where GHQ India would be in control of Iraq and Persia.[36] Indian Army procedures at the time were to wait until demands were made on India for a contribution. Then WO approval would be secured. Later on, material would be gathered or soldiers trained. This long, drawn-out practice took anywhere from six to nine months. India's transport contribution to the war effort by this time was thus meagre. She had only raised one company of transport troops, trained at the (Royal) IE Depot at Jullundur, which had been sent to East Africa, although several other railway companies were in the process of being raised. There was local senior military resistance to India taking on a greater share of the war's burden before Australia, South Africa and other colonies had done their part. When demands came from the Middle East, Indian authorities felt these to be outside of their area of concern. On 8 April 1941, the demand of 100 more miles of track for the Middle East, for example, met with resistance by the EinC and the QMG because there would perhaps be political repercussions in India and this would reduce travel capacity in India. At the same time, the colonial government in Burma rejected calls to supply transportation troops and material because Burma was separate from India and indeed had 'nothing to do with India'.[37] Nor could the Director of Transportation at GHQ India visit Burma to investigate the possibility of setting up a Transportation Directorate, as Burma was an independent command. A follow-up effort to obtain an invitation from Burma to send a transport officer there for liaison purposes was also rejected.[38] India would have to shoulder the weight of transport support in the region on its own.

Although it was recognized that changes had to be made in the administration of transport in Iraq, this resulted in no change in policy in India. Both the EinC and the QMG resisted any such change, and India did not alter the procedures of waiting to be called upon to develop new military transport resources. This was due, Biddulph argued, to a 'lack of knowledge and experience [that] limited appreciation of the world situation'.[39] One of his first priorities was to change the way in which operating units were mobilized. The prevailing system at the time in India was that no move would be made until a theatre of war put in a request for a unit to the WO. If the WO approved, one would be organized, which meant many months of delays and India not contributing as much as it could to the Middle East or Far East. Instead, the Transportation Directorate would be proactive, engaged in advance exploratory work and put in demands on the basis of projected need and thus organize operating units before they were needed, quickening response time and expanding India's transportation remit in other theatres of the war.[40]

Some steps had already been taken to train engineers. In February 1941, India had formed the No. 2 Transportation Depot, IE, at Jullundur, with a complement of forty-three men. No. 2 Transportation Depot, IE, had the responsibility to collect personnel recruited by either the Calcutta Port Trust or the railways and to form these men into Docks and Railway units. The numbers produced, however, remained too small; between the beginning of the war and May 1941 only 1,000 transport troops had been raised at Jullundur.[41] A memorandum was thus issued on the provision of transport units that includes proposals to raise 27,000 transport troops who would form new units within eighteen months. The Indian General Staff agreed immediately to the proposals. However, the QMG objected because these numbers were excessive. This was overcome by bringing to QMG Branch's attention Hall's *Inland Water Transport* in Mesopotamia, The Royal Commission's *Report on Mesopotamia of 1917* and *The History of Transportation on the Western Front*, all texts QMG Branch had not previously examined. Although Jullundur had successfully formed eleven such transport units, it was decided that the new, higher demands and the need to specialize training demanded separate training facilities. Thus, on 1 July 1941 the No. 5 Engineer Depot, IE was created to raise and train IWT and Docks units, some staff being transferred from No. 2, which now focused on railwaymen only. Between May and mid-September, a period of four months, the training of transport troops was dramatically expanded, and the number of trained transport personnel increased from 1,000 to 40,000 men, including reserves, reinforcements and wastage.[42]

Biddulph was concerned by July about the unravelling transportation situation in Iraq and the problems of relying upon civilian transportation staff in a conflict area. Transportation in Iraq in support of the army remained at this time under civilian management, with minimal interference from the military. India had been gathering significant amounts of locomotives, rolling stock, rails, cranes and other transportation material for Iraq, and these were beginning to be despatched.[43] Unfortunately, the rebellion that broke out in Iraq revealed the inadequacy of the transportation arrangements in the country. Civil labour stopped working on the Basra Docks and the railways ceased moving.

To rectify the situation a new transportation infrastructure had to be developed to support a larger military force. India now sent locomotives, wagons and transportation personnel. This began with the 'Greasers Wire': a telegram requesting 200 men skilled in the trades, including firemen and greasers, who could work the essential river services. As civilians were sent (and they remained un-militarized) compulsion could not be used. These civilians were viewed as a source of continual trouble by the military, beginning with an initial strike and refusal to embark at Bombay until they received more pay. The railways were also a problem, as senior railway officers used the limited railway infrastructure that had existed in Iraq in the past as an excuse not to send more material. Without considering ongoing and planned railway expansion, in one case a request for 300 wagons was criticized, as it would only add to congestion on Iraq's lines. The QMG hesitated to make judgements because of contradictory advice on such matters.[44]

Biddulph hoped Lt. Gen. Wilmot Gordon Hilton 'Phil' Vickers (1890–1987), recently appointed as the QMG in Iraq, would militarize the civilian transport staff and resolve the problems.[45] Nevertheless, in his view, Iraq also needed an in-country Director of Transportation. Brig. Sir F. Carson had indeed been appointed as Director of Transportation for Iraq but he took considerable time to come to the Theatre and when he arrived in India, he went on sick leave, delaying transportation development in Iraq as India's policy was not to send out men or material until a demand was made. Demands for different kinds of transportation material had to be directed to firms in India. The correct procedure for this was for demands to be made to the Chief Engineer, Iraq, who would communicate the demands to the EinC, India, who would communicate the demand to the correct firm after consultation with the Director of Transportation, GHQ India. If there were questions, they would have to go to the director and settled there or he would direct these back through the chain of communication to the Chief Engineer, Iraq. Unfortunately, the Chief Engineer, Iraq and the EinC in India disagreed about how to obtain necessary supplies, with the result that no transport material was moving to Iraq and there was yet no director on the spot in Iraq to help facilitate demands.[46]

Biddulph's main concerns were two-fold. First, Transportation was lagging far behind required demands, at a time when it was only beginning to make headway in India, and someone had to go to Iraq urgently to fix the situation before the transportation organization failed altogether. Secondly, the British military presence in Iraq would be difficult to maintain if transport support did not work. If the military could not be sustained in northern Iraq, it would be difficult to hold onto Basra. Biddulph thus made several requests to visit Iraq until this was approved in August 1941 so that he could act as Director of Transportation in Iraq only as a temporary measure until Carson could take up the responsibilities. Carson would recover and report for duty on 16 October. Nevertheless, Biddulph was ordered not to return to India but to remain in Iraq, now as Deputy Director of Transportation in Iraq under Carson.[47]

While Biddulph was in Iraq, Manton arrived in India on 11 October. Manton was now appointed brigadier and officially became India's first full Director of Transportation (October 1941–June 1942) at GHQ in Biddulph's place. Biddulph argued later that this change in the direction of the development of the Transportation Directorate removed continuity at a crucial point and many plans still embryonic when he left never materialized because of the change in leadership.[48] However much that might have been true of some programmes, the major ones most directly involved with railway work continued on their course or were further expanded under Manton. Manton, for example, found that the engineer depots were not producing sufficient numbers of transportation men. To further increase the number of men available for the railway units, the civil railways undertook to form Technical Training Groups to provide instruction in railway trades, and later, collective training in operating a railway as a unit was provided by using several nugatory branch lines. To provide military training to these men, Manton set up two Transportation Training Centres (TTCs) in October 1941. No. 5 Depot, IE, becoming No. 1 TTC at Bombay and Deolali now devoted to the training of docks and IWT personnel. No. 2 Engineer Depot at Jullundur now became the No. 2 TTC devoted exclusively to train railway personnel. At first the curriculum focused on giving military training to civilian railwaymen recruited in India to form railway operating units, but over time, the centre undertook to provided technical training as well as well as refresher courses. The staff of No. 2 TTC would increase over the course of the war from 43 men to 333.[49]

Aiding Manton in India was an Irish engineer and son of a British railwayman, William Hillary Prendergast (1895–1957). Prendergast was crucial because he had recent imperial experience with railways. The first occasion came in the First World War when, after being commissioned in the Royal Engineers in 1916, he served in Mesopotamia where he remained until 1918. He joined the Bengal-Assam Railway in 1920 (which would remain his official employer in India until his retirement in 1947). His tasks were not limited to normal railway operation and maintenance of the permanent way, signalling systems and buildings, but included design, organization and execution of various new railway projects. It was this work that led to his appointment as Deputy General Manager of the railway in 1938 and his deputation in 1939 to Britain to investigate how the home railways were organized. The intent was for Prendergast to copy British home railway management and implement them on the Bengal-Assam Railway so far as was possible. This research consumed six months in 1939 as Britain entered the Second World War. When Prendergast returned, he applied what he had learned to the Bengal-Assam Railway's operations 'with excellent results'. The railways lent Prendergast to the British Army in 1940 and he was commissioned as a captain (temporary major) but then made an acting lieutenant colonel of the IE. Prendergast was then sent in 1941 to Iraq where he planned and implemented the doubling of the railway line from Basra to Baghdad and where, presumably, he worked with Biddulph.[50]

One of Prendergast's main tasks was to raise and train railway construction companies.[51] One such unit, the No. 3 Railway Construction & Maintenance Group, organized at Jullundur out of such men in November and December 1941, would play a prominent role in railway operations in Burma during the Japanese invasion. Prendergast would also take this unit into Burma as the Officer Commanding in 1942. Before this deployment, very little military training was provided to such units. As Prendergast commented in his report on the campaign after he left Burma in May 1942, for example, his adjutant, Capt. R. A. Busher, had to give the men the necessary training in using weapons 'on the fly' while in Burma over the period covering February to May.[52]

Chapter conclusion

The adaptation of military reforms is not an easy or an even process across different forms of knowledge, military organizations and personalities. As has been discussed in this chapter, military Transportation Technique did not command assent to its adaptation on its virtues alone. Indeed, even with vigorous championing by most expert people responsible for the railways, its acceptance was slowed by ignorance, interest groups and converse priorities. The heads of civilian transport agencies in India wanted metropolitan military interference no more than the commanders of the Q Branch were aware of its benefits or indeed what it actually was. While everyone would readily admit that gas masks should be adopted in the face of gas attacks and that wounded men needed bandaging, evacuation from the field of battle and penicillin, the proper procedures for operating trains and organizing Transportation Technique (much less Movement Control) was seemingly both too abstract and too mundane to capture attention.

The reformers who tried to encourage adaptation of military Transportation Technique were much more committed, would persevere and would in the end be successful in the acceptance of the Technique but only after significant damage to their own reputations and even careers and only after delays in its implementation had indirectly caused the deaths of thousands. This struggle and these vicissitudes will be the subject of the chapters that follow.

2

Local, colonial railway experience

Many colonial railwaymen, whether in Burma, the Gold Coast (Ghana), Sri Lanka or any other local railway in the extra-European world, sincerely disliked or did not respect outsiders with little in-country experience playing a role in railway operations or management. The BR men in place at the time of the outbreak of the war, most having worked in Burma from the early 1920s, were secure in their authority over the railway and in their career paths. This remained the case despite Burmanization, because although they were white or Eurasian, they had seniority and so while new vacancies would go to Burmans, the white and Eurasian railwaymen hoped that they would keep their own jobs until retirement. Nevertheless, they realized that their positions and prospects were more vulnerable to transgression by white authorities from other colonies, but only if their own local expertise counted for nothing.

BR men's concerns about outside authorities arriving to take charge, knowing little of Burma and the limitations of and normal demands on the railway may have been genuine. However, it was their continued subversion to the control and direction of officers imported from outside that some would dislike. Biddulph's arrival was very threatening to the BR men, because military management showed little concern for the local experience that Burma's railwaymen relied upon for their authority over railway operations and management. Although these railwaymen would voice their discontent as a problem between the military and civilian ways of doing things, it was really the management of the railway by an outsider imported from the home country per se that caused friction between Biddulph and the BR men in 1942. Often this hostility undermined railway operations designed to save the colony and the railwaymen themselves along with everyone else, from capture by or death at the hands of the Japanese Army. At times, this hostility was even fever-pitched. Given the crisis in which they found themselves, the unwillingness by many railwaymen to cooperate with military authorities seems extreme.

The present chapter examines why these railwaymen were so resistant to outside interference, why men who would evince so much group loyalty would be among each other's worst professional rivals and how tensions and bonding among them, seemingly contradictory passions, both became especially powerful on the eve of the Japanese invasion. These contradictions were a product of an era of indigenization (or Burmanization as it was called in Burma) of colonial bureaucracies. In the case of the BR, many white, Indian and Anglo-Burman railway officers felt secure by relying more heavily upon their local expertise in railway operations in the local colonial context for their continued hold onto their social position, their occupational status and their authority over railway operating. In the 1930s, a new threat from Depression-era retrenchment efforts joined a colonial political climate that from the 1920s had witnessed the gradual erosion of a guaranteed claim to race-based social status. Retrenchment made insecurities among railwaymen worse by threatening chances for professional advancement, and for particular individuals, these prospects seemed bleaker still because of a new administrative scheme introduced just prior to the outbreak of the war.

Local experience and rivalry

The ranks of the railwaymen remained very heavily skewed towards 'Europeans', which was used to categorize all 'whites' regardless of geographical origin, at the highest levels, making up two-thirds of gazetted officers in 1940.[1] This category consisted of a small cadre of railwaymen drawn from around the empire, but mainly from Britain, who managed, attended and operated the railways, its locomotives, stations and workshops. Being white gained for these railwaymen employment in colonies like Burma in the nineteenth and early twentieth centuries, but colonial policies across Southeast Asia had begun to work against white civilian employees, less so the military officer ranks, after the First World War. However much racial identification remained a very potent factor in guaranteeing their status and employment, it was very clear that their pedestal was being chipped from below. 'Racial' data for railway employees at the end of the 1930s, for example, indicates a small but steady increase in the percentage of non-European officers and staff among railway personnel in Burma (Table 2.1).

The railwaymen's position in Burma was mainly buoyed by their local experience. Like most colonial railwaymen, those in Burma were a 'mixed bag' of men with varying levels and kinds of technical, occupational or professional

Table 2.1 Racial divisions among Burma Railways staff in 1940

	Gazetted Officers	Non-Gazetted Employees	Total
Europeans	58	16	74
Anglo-Burmans	11	582	593
Burmans	16	4,916	4,932
Indians	6	16,010	16,016
Others	–	286	286
Total	91	21,810	21,901

Source: PRO/WO/106/2654 Enclosure 28, the Report on Burma Communications, p. 4.

training. Only a few had formal training or apprenticed in railway work outside the colony. Although many had held previous jobs in Britain, India, Australia or elsewhere in the empire before joining the BR, this was usually in other kinds of work. Some entered the BR's service straight out of school, such as Australian Thomas C. Parker who left Perth to join the railway as an assistant auditor in 1919. Only a few had general formal training (or had apprenticed) in railway work, including how to steam up an engine, shunt a train, recognize signals and so on. Some BR men could boast of extensive railway experience before coming to Burma. One was Norman Johnson who rose to become Chief Mechanical Engineer for the railway by 1941 and who would be one of the 'Last Ditchers' in Rangoon in 1942. Johnson had originally apprenticed at the Swindon Works of the Great Western Railway before and after the First World War before the North Western Railway of India hired him in 1920. He only transferred to the BR in 1932.[2] There were also those with previous railway experience from service in one capacity or another in the First World War. Most of the BR men, however, acquired their professional knowledge in railway work in the colony.

This book considers the knowledge that these men gained about railway operating in the Burmese context to be *local expertise*. *Local*, especially when deployed in opposition to *metropolitan* or *imperial*, has been used to mean different things in several disciplines, making it difficult to deploy for the purposes of this book without clarification. The pursuit of more accurate and bias-neutral terms to describe their objects of study has led anthropologists to use *the local* today as they might have in the past used *the savage*, *the tribe*, *the native* and *the traditional*.[3] In the anthropological and the developmental studies literature, *local knowledge* is usually taken to be non-Western knowledge.[4] Scholarship in the history and sociology of science often uses *local* in this way too, although it more commonly uses *local* as a binary opposite to *metropolitan* as a locational referent to distinguish the main centres of scientific research and that research and discovery that occur on the periphery, a difference of status and resources, rather than of autonomy. The particular subfield of the history of science in the colonial world, however, has frequently used *local* in an anthropological way and hence uses *indigenous*, *traditional* and *local knowledge* interchangeably, as non-Western knowledge that has been challenged by the arrival of Western knowledge.[5] A third way in which *local* is used is by scholarship that deals with the impact of state planning, in which it can mean the hands-on experience developed in the particular circumstances of a locality as opposed to central or national plans that emerged out of formal scientific or technical learning, that are then applied generically as 'one-size-fits all' plans. This kind of local expertise is best captured in James C. Scott's idea of informal, bespoke, hands-on knowledge.[6] Scott was making a contrast to high-modernist and bureaucratic city planning that led to the failure of a number of state projects, with Scott targeting 'imperial and hegemonic planning mentality' for criticism.[7]

This book takes local expertise to mean technical knowledge and knowledge essential for technical operations as informed by the local context. In other words, local expertise developed on the basis of experience in the locality, locally positioned technical men being understood to be part of the same technical universe as planners at the centre, but having developed a new perspective on it, at least relative to the congruence of externally developed technical knowledge with the local context, giving them valuable local expertise. As Aaron Moore shows, this kind of local expertise is reflected in the opposition of local engineers to state planners in pre-war Manchukuo.[8] Nevertheless, it was also indispensable to successful railway operating in the colonial context. Feelings were probably strongest about the importance of local expertise where contingents of colonial railwaymen remained small and where their individual experiences counted for more and gave them more authority over railway operations than they might have found in more densely populated railway staffs. In a colonial setting, twenty years and more of 'in-country', locally derived experience provided the white railwaymen with an almost unassailable expertise in local, colonial railway operations.

The knowledge most of these men had of operating trains was learned in Burma. This made their trade a craft for experience was its only source. There were so many challenges to running a railway in a colony like Burma, and other railwaymen would certainly have felt the same way about their operating environment that took years of hands-on experience to develop skills with which to cope. This included such things as knowing how to find water, a skill that would be useful in Burma since railway operations on several branches passed through dry country. The threat of torrential downpours required special caution during the rainy season. At higher levels of management, the railway's technical aspect, as we shall see in later chapters, was of crucial importance to understanding the railway's limits, just as understanding the nature of traffic on the line, rather than simply the volume, was essential to knowing what demands would be put on the railway and when.

To some degree, this kind of local expertise overlapped with what E. P. Anderson referred to as the 'railway instinct'. A few courses alone did not impart it. A railwayman acquired this instinct through regular experience in railway operations and it gave him the ability to make the right choice in an emergency. This instinct was also essential for railway operations in a war context, wherein emergencies were usually commonplace. As Anderson explained,

> The 'railway instinct' is something not unlike the sea instinct, which takes years to acquire but which makes men who possess it do the right thing in an emergency almost or quite automatically ... Officers and men cannot keep their railway instinct alive and keep up their technical skill and knowledge unless constantly making use of them ... So we are forced to rely on being able to secure on mobilization the services of those officers and men whose work in peacetime qualifies them for the duties required of them in war.[9]

Local expertise also included cultural knowledge that was potentially relevant to railway operations. The future president of the United States, Herbert Hoover, who built his fortune as a young engineer after co-founding the Burma Mining Company (later, the Burma Corporation) in Upper Burma, for example, wrote about this in his memoirs. Hoover recalled the story of one of the European bridge builders whom his mining operation hired to build the Burma Mining Railway connecting the mines to the main line at Namyao Junction on the Lashio Branch. The Burma Mining Railway had been having vexatious troubles in its attempts to use Shan engine drivers to work the railway before Europeans or white American engineers could be hired. These drivers believed that *nats*, malevolent spirits, were unhappy with the obstruction of the river by the bridge. Shan bridges were suspension bridges that allowed *nats* to pass under freely. The present railway bridge, by contrast, blocked the river gorge with the girders that supported it. Thus, the Shan drivers would not take the engines across. The aforementioned young bridge builder fell while walking carelessly on the bridge, smashing into the girders on the way down to the river below, breaking ribs, nearly drowning and otherwise appearing very dead until he was resuscitated and taken away to hospital. The manager put up signs claiming the man had been sacrificed to the *nats* and they had now been propitiated. The Shan drivers accepted this and now took the engines across freely.[10] This story indicates some of the nuanced knowledge of indigenous religion and culture that railwaymen in different colonies had to be aware of because it could interfere with railway operations. Local expertise also included the command of languages, for the BR men were generally required to pass a 'native language' test. William J. Air (the Deputy Railway Commissioner on the outbreak of the war) and Charles Patrick Brewitt (the Deputy Traffic Manager before the war and soon-to-be head of the Railway Operating Company during the FBC) were fluent in both Burmese and Hindi (at the time termed 'Hindustani'). Other BR officers were also tri-lingual and passed language examinations usually in different combinations of Burmese and Hindi,[11] or, more rarely, Burmese and Urdu, another Indian language, as in the case of P. Manifold.[12]

The importance of local expertise was intertwined with work-related corporate identity and group cohesion. In pre-war Britain, where many (but again, not all) of the BR men had their first experiences with railway employment, private railway companies encouraged railwaymen within the company to think of themselves as a corporate body and not as individuals. As Frank McKenna has suggested, this sense of corporate identity was strengthened by the high status in the community that railway workers enjoyed in the days of steam. There was also status by degrees among the company railwaymen recognized by rewards, ranking and special pay. No industry in Britain, McKenna asserts, ever before or since had created the corporate spirit as had the private railway companies and it was this corporate spirit among their workers that 'marked them off so clearly from other complex organizations'. The same assertion could be made of the BR before 1942.[13] Cohesion was also the unconscious result of working within the same organization and among the same men for years. According to Joseph A. Olmstead, over the course of their working together, organizational incumbents will develop 'norms, status structures, and patterns of interaction' that will shape how members approach their work tasks and their motivation in achieving these tasks, making them a team. Such team qualities are strongest in primary groups such as platoons or gun crews, to cite examples from the military, but there is potential for the development of such qualities above the primary group as well.[14] No matter what their personal feelings towards each other, and as we shall see these were often not positive, the sociology of the shared work experience of a relatively small group of white men in a colonial setting defined by race at the level of skilled railway operation set the conditions for team cohesion.

The BR men also worked together very well as a team in the trains, in the yards and across the network because of the peculiar nature of the railway as an organization. From the 1830s, the railways in Britain, on which the three systems that eventually made up the BR were modelled, had themselves used the military as their model for terminology and techniques on how to control railway labour. As McKenna explains, when the size of railway companies mushroomed in Britain, the only large-scale organization available that could serve as a precedent was the British military. A strong service ethic was developed and reinforced with slogans that conjured up images of military posters during wartime, stressing loyalty, hard work and efficiency. Railway manuals provided for a graded hierarchy that structured the railway labour force. Railwaymen were considered to be 'in service' with a range of restrictions on leave and legally enforceable responsibility to duty.[15]

Familiarity with the military way of doing things led demobilized officers and soldiers into BR service in the first place. Their military experience made them suitable candidates for railway service even when their military experience did not include railway work. It is true that after the FBC, Governor Reginald Dorman-Smith suggested one reason for the failure of militarization was that BR men 'cordially disliked military control'.[16] This assertion does not hold up to careful scrutiny of the available evidence as will be shown in the following chapters. For the moment, a brief glance at some of the BR men gives some indication of the kinds of military experience they brought with them. E. V. M. Powell had done technical training at Crewe Mechanics Institution before the First World War broke out, then drove trucks at Woolwich in the Arsenal Transport Company before doing the same for the Serbian Army and then in Salonika. Powell followed this with depot work in Constantinople and a brief stint with the London and North West Railway before joining the BR in 1920.[17] Ernest Proctor had been a civil engineer just before and just after the First World War, but during the war he served in the British Army, before being hired by the BR in 1920 as an assistant engineer.[18] William James Air, serving as both the Deputy Commissioner of the BR and the Secretary to the BR Board in 1941, had entered the First World War as an artilleryman.[19] Some railwaymen had only military officer experience, admittedly an experience that prepares one for managerial demands. Brewitt, who would rise to the top of the militarized railways in Burma during the FBC, had no technical training before the First World War. During the war, he served in the British Army in operations in Salonika and

Serbia until 1919. By that time, he had become a lieutenant and was awarded the Order of the British Empire (Military Branch), before joining the BR in 1920.[20] What these men found in the BR was the same kind of emphasis on teamwork and group cohesion they had become familiar with in the British Army.

Nevertheless, there were some factors that worked against team cohesion among the railwaymen. Before the war, there were examples of personal rivalries within the BR men's ranks. Opportunities for promotion and better pay were also a part of this equation, for the highest positions were the prizes of years of career building and planning and such considerations could and did damage the cohesion of the BR men as a working group. After separation from India, the BR were put under the BR Board, a statutory corporation, headed by a British officer who served as the Chief Railway Commissioner (hereafter, CRC) or the *Mi-yahta Min-gyi* (great lord of the fire carriage) to the Burmese. Outside of the Board, the administration was officially structured down from the CRC.[21]

It took some individuals several decades of institutional networking, infighting and even backstabbing to build their careers. How one moved up or down the railway hierarchy depended in part not only on seniority but also on appraisals by higher-ranking railway administrators. Although there was no shortage of adulators beneath them, there were also clearly a number of lower ranking members of the railway staff who had had entanglements with those above them. An extensive examination of the memoirs of high-level BR officials in place in 1942 and before, official records and confidential appraisals of staff contained in the BR Department's personnel files paints a picture of deeply bitter personal rivalries between several of the railway officials who would play key roles in the railways in the FBC. Instructions in the footer of the evaluation forms that were produced at some point between the end of the FBC in May and October 1942, when a series of confidential reports on BR personnel were produced, mandated administrators to be frank and very personal about their subjects, highlighting anything that might affect the judgement or work of these men.[22] The reports are frank and provide insights into personal rivalries, anxieties, jealousies and other examples of personal animus.

Normally, such problems did not affect railway operations, at least in any discernible way from the perspective of the records today. Perhaps one of the main reasons for this was the geographic spread of railway offices which, when combined with deployment strategies, provided a safety valve of sorts for dealing with interpersonal problems by keeping railwaymen who did not like each other as far apart as possible within the railway network. The colonial

BR, like most of the pre-1948 railways in Britain,[23] applied the departmental rather than the divisional system. The departmental system was better suited to a smaller than to a larger railway system and so tended to dominate the organization of early railway systems. This was a highly centralized system, a kind of administrative panopticon, in which the railway was divided by function across the whole network, allowing network-wide view throughout the whole system from a control office in the railway headquarters (HQ). These could include traffic, mechanics, civil engineering, accounts, construction and other departments. Each department was broken up into sections or districts that did not necessarily overlap or agree in number. Each district was also under a district chief who was supervised at a central level by their department chief who answered to the chief manager of the railway.[24] One historian has argued that in the case of the railways in the United States in the nineteenth century, this kind of vertical system prevented young officers from developing personal initiative as they instead merely sought to win superiors' favour.[25] The departmental system also meant that decisions tended to be better informed regarding the railways as a whole but with a weaker grasp of the local situation.[26]

By contrast, the divisional system, which began to appear among India's railways from the 1920s, was often better in a larger railway system where governing all aspects of railway operations from a central office was challenging, especially given communications limitations of the pre-war era. In this more horizontal system, each territorial division of the railway was a separate sphere controlling all aspects of railway operations within it. Each division would be under a single officer who would have total responsibility for the operation and running of the railway, the status of the permanent way and the rolling stock. Such a system was more responsive to local situations. However, as one scholar has argued, it also spurred huge rivalries between division heads who sought to promote their own division at the expense of the performance of the rest of the network.[27]

There were other manifestations of these competing systems on different scales and so many systems were in fact mixed systems in both organization and practice.[28] Ralph L. Wedgwood, who later headed the Wedgwood Committee in India in the mid-1930s, was a delegate to the International Railway Congress held in Washington, DC, in May 1905. There, he offered his views in a paper entitled 'Need for Reform in Railway Organization', which was based on his experiences when he was the Divisional Goods Manager for the North-Eastern Railway at Newcastle. Wedgwood saw the differences between the divisional system, which he described as the American system, and the departmental system, which he referred to as the British system, as a difference in the scale responsibility of individual superintendence. Wedgwood explained that the main problem identified with the divisional system was that it put too much responsibility for different processes within train operations into the hands of one man, the divisional superintendent. However, the departmental system, his own, was far worse as consultation was needed on each issue of train working, consuming time.[29] Wedgwood also argued that the divisional system offered other advantages. First, this system gave railwaymen a certain sense of solidarity. It allowed a divisional officer to 'infuse energy into all ranks of staff' which he could not do in the departmental system. In the latter, two-thirds of staff (e.g. in a three-department system) were responsible for other departments and were outside the officer's authority and influence. The divisional system, Wedgwood argued, also 'open[ed] a better career to railwaymen by ensuring that in the operating department they will obtain a thorough all round training in every branch of operating work'.[30]

Nevertheless, a divisional system as understood at the time was not the panacea that Wedgwood seemed to suggest to all of the problems facing railway operations. The divisional system would have to evolve over time as the twentieth century progressed and would see greater subdivision and specialization within each division, producing in effect a mixed system that sought to include the advantages offered by both systems.[31] The BR would attempt to adjust its own departmental system in the colonial period, but would eventually make the change to a divisional system in the post-war period.

Regardless of which system the BR might have opted for, the wide geographical spread of the BR meant that staff were widely scattered. Aside from the uppermost tiers of departments, usually consisting of a few individual and support staff, the departments themselves were subdivided into numerous divisional offices spread throughout the railway network. Often these were located at a vast geographical distance from the railway HQ in Rangoon. This meant that white, Anglo-Burman/Anglo-Indian and Indian middle managers of the railways were not concentrated in Rangoon, but in a number of departmental centres. The BR afforded much space for personal distance and just as one could seek transfer to one of several of Burma's cities and major towns, a sufficiently powerful middle manager could find a way to send a disliked colleague to some distant outpost in the system, or simply to keep them out of the railway HQ in Rangoon (Table 2.2).

Table 2.2 Geographical structure of the Burma Railways administration – HQ of departmental districts, circa 1941

Engineering Department	Locomotive Department	Traffic Department	Signal Department	Medical Department
Rangoon	Rangoon	Rangoon	Rangoon	Rangoon
Toungoo	Toungoo	Toungoo	Toungoo	Toungoo
Mandalay	Mandalay	Mandalay	Thazi	Mandalay
Pegu		Pegu	Mokpalin	Insein
Ywataung		Ywataung		
		Henzada		

Source: PRO/WO/106/2654 Enclosure 28, the Report on Burma Communications, p. 4.

These offices were scattered about the railway network so that they maximized the availability of administrative and operational resources to as many branch lines as possible and hence tended to be concentrated in major junctions. The operating sections and branches of the BR were referred to by varying names, depending on the period, context and who was speaking. Official reports tended to refer to sections and branches according to the termination points, for example, the Mandalay-Myitkyina Branch Line, or Myitkyina Branch for short. Nevertheless, railwaymen themselves often used older names, inherited from the pre-unification days (i.e. before the creation of the BR when this or another branch was an independent railway). For these men, the Mandalay-Myitkyina Line was the Mu Valley Line. For consistency and convenience, the present book refers to the lines and branches according to the shortened version of the official usage, as shown in Table 2.3.

Pre-war tumult

Particularly serious compromises to the position of senior white railwaymen in Burma were posed by the expected effects of changes in colonial railway policies sketched out just before the war. These new policies were directed at dealing with road competition that shook up the ranks of the railwaymen and the internal institutional cliques that secured one's position within the railway hierarchy. Max Weber's concept of the appropriation of social and economic opportunities monopolized by a particular group is particularly useful. In this model, outsiders are not the only ones closed off to access to the monopolized

Table 2.3 Section and branch line names of the Burma Railways, circa 1941

Formal Name		**Other (or Original) Names**
Short	**Long**	
Bassein Branch	Letpadan-Bassein Branch Line	NA
Burma Mines Railway	Burma Mining Corporation Railway	NA [Note: Private Line]
Katha Branch	Katha-Naba Branch Line	NA
Kyaukpadaung Branch	Pyinmana-Kyaukpadaung Branch Line	NA
Lashio Branch	Mandalay-Maymyo-Lashio Branch Line	NA
Madaya Railway	Mandalay-Madaya Light Railway	NA
Martaban Branch	Pegu-Martaban Branch Line	NA
Mingaladon Branch	Mingaladon Cantonment Branch Line	NA
Myanaung Branch	Henzada-Myanaung Branch Line	NA
Myingyan Branch	Thazi-Myingyan-Mandalay Branch Line	NA
Myitkyina Branch	Mandalay-Myitkyina Branch Line	Mu Valley State Railway
Prome Branch	Rangoon-Prome Branch Line	Rangoon and Irrawaddy State Railway
Rangoon-Mandalay Trunk Line	Rangoon-Mandalay Trunk Line	Sittang Valley State Railway
Shwenyaung Branch	Thazi-Shwenyaung Branch Line	Southern Shan States Railway
Thongwa Branch	Pegu-Thongwa Branch Line	NA
Ye Branch	Moulmein-Ye Section	NA
Ye-U Branch	Mandalay-Ye-U Section	NA

Note: NA = Not applicable.

Source: Collated by author from data in the *Burma Railways Reports, 1937–1941.*

resource(s), for there are also varying limits to access to this resource within the insider's group.[32] We will see in this section and, indeed, in the chapters that follow a struggle among the insiders, the white railwaymen, for positions at the top of the railway hierarchy. Going beyond Weber, however, some insiders would ally with outsiders, in the form of imperial railway authorities in 1942, to gain control of the monopolized resource for themselves.

This shake-up had to do with roads. A lack of colonial investment in roads, in part because of the BR's pre-1937 opposition to anything that would encourage competition from motor traffic, meant that Burma's road network was poorly developed into the early 1930s. For much of the colonial period, many of the roads were often simply feeder roads for the railways or river traffic. As highways were constructed in the 1930s, however, there was increasing competition with the railways that began to dig deep into railway revenues as passenger traffic earnings began to decline after a 1929 peak. The railways both in Burma and in India were worried. The railways thus began to undertake measures to improve performance to attract more customers, including speeding the rate of transport between cities and towns.[33]

A bold restructuring of the BR was recommended after a comprehensive investigation conducted in 1936–7 by the Indian Railways (IR) Enquiry Committee (the Wedgwood Committee), which arrived in India in November 1936. After years of losses, in particular from increasing competition from road carriers, the Wedgwood Committee was tasked with examining the state of the IR and making suggestions regarding better coordination of rail and motor traffic so that they did not drive each other under and to improve railway profitability. The Wedgwood Committee spent several months on tours of India, Burma not being visited, and then issued a report in 1937.[34] The Wedgwood Committee was highly critical of the IR's determination to be at the forefront of modernity. The railways in India, just as it had in Burma, prided itself in being the vehicle of Western civilization and modern technology to the Indian and Burmese masses. The railways were a virtual advertisement for colonial success and imperial legitimacy. By bringing the railways, the British made life for the indigenous population better and this made India and Burma prosperous. Keeping up appearances was thus a vital response to the economic problems of the 1930s and the IR and the BR were building the stations, workshops and marshalling yards that were 'the last word in railway technique', but exceeded requirements. While railways in Europe were exercising austerity, the IR and the BR were pushing the limits of what was financially feasible in the pursuit of prestige.[35]

The Wedgwood Committee Report also promoted British austerity and ingenuity for improving profitability in even the worst of times as the right attitude for India (and Burma).[36] Although at the time of the issuance of the report the BR had already separated from the IR, the BR Board took its recommendations to heart. In some cases, state-of-the-art technical improvements in monitoring railway traffic were pared down and some smaller line extensions and branches that were not yet showing profit were abandoned, changes that meant little at the time, when measured in terms of railway profits, but would to some degree impair railway performance when the Japanese invaded in 1942. More serious, the BR Board used the Wedgwood Committee Report's recommendations to justify the resurrection of the Organization Department that had been defunct since 1934, giving it a mandate to increase railway earnings and save the railways money.[37] The Organization Department then set about making changes in the district arrangement of the railways that reduced the number of other senior posts.[38] This eliminated availability in four senior postings, including two district engineer-ships (although one of these had been inherited from the previous year), one district traffic superintendent-ship and one district locomotive superintendent-ship. The CRC of the BR, John Rowland, ordered an investigation that revealed how these changes had negatively impacted the career expectations of some of the existing white railway staff. This concluded that while the financial damage could not be easily calculated, it was clear that many railwaymen had experienced damage to their careers.[39]

Some changes were also promised for the Head Office. On the basis of the Wedgwood Committee Report, which had also called for giving the commercial and operating side of the Traffic Department equal status, the BR decided to pull the commercial side of railway operations out of the hands of the Traffic Department and make it a separate entity as the Commercial Department.[40] The strengthening of the prospective Commercial Department required changes in senior posts. It would require the removal of a senior administrative post and the raising in status of two junior posts to senior administrative posts by 1938.[41] Another, perhaps more unsavoury aspect of this change for those junior railway staff who were in line for promotion before 1938 were the special conditions now put in place for the new posts, at least for the commercial side. In the past, all senior posts were open to anyone who had reached the level of district traffic superintendent. These in turn were open to those who had been assistant traffic superintendents, so that once someone was hired by the railways and survived their probationary period they could expect after a certain number of years of adequate service to achieve a senior post before retirement. This arrangement would now be changed. The new commercial postings, taken as a model perhaps of what was ahead for the redefinition of other senior posts in the future, would be only open to those with specialized training and experience, which meant perhaps, these posts might even go to outsiders if the local talent did not possess the formal qualifications for the job. At the very least, several senior postings were now closed to junior administrative officers of some years' experience who had not already held several specific posts earlier in their careers.[42]

The fact that the position of Traffic Manager was vacant in 1941 would have significant consequences for the railway administration during the evacuation and an enormous influence on the source material for understanding railway performance during the period. Two men were in position to accede to this post. E. I. Milne and Charles Patrick Brewitt had both joined the BR in 1920, both in the Traffic Department, and both at the same rate of pay. Both became District Traffic Superintendents in the early 1930s; although Milne would remain so well into 1939, Brewitt followed the administrative success of John Rowland. When Rowland became Agent, BR (changed to the position of CRC with the separation of the BR in April 1937), Brewitt was promoted to officiate in his first junior administrative post from 9 April 1935, when he became Deputy Agent, and later as Deputy Railway Commissioner, until 31 August 1939. On 20 April 1940, Brewitt was appointed to the post of Deputy Traffic Manager (Transportation), another junior administrative post, which he would occupy until 24 February 1942.[43] Milne, who had been languishing outside of the Head Office far behind Brewitt, was finally made Commercial Research Officer in 1939, which soon after changed to Deputy Traffic Manager (Commercial), as a result of the efforts by the BR to introduce the recommendations of the Wedgwood Committee Report. Brewitt had benefitted from the several years that his mentor, Rowland, had been Agent and for the four years (from April 1937 until February 1941) that Rowland spent as both CRC and President of the BR Board. That dual position gave Rowland a particularly strong say in both railway policy and daily administrative decisions, including those related to personnel and advancement.[44]

The Rowland era seemingly ended in February 1941, with negative consequences for Brewitt. On 28 February, the grand old man of the BR stepped down as both President of the Railway Board and as CRC of the BR as he took furlough preparatory to retirement in Australia. A knighthood in the same year honoured his contribution to the development of the BR and henceforth he was referred to as Sir John Rowland. William Henry Chance now assumed both positions, giving him the clout to help friends and hinder enemies.[45] Milne now followed the good fortune of Mr Chance, who became CRC in 1941 when Rowland retired. Of the two Deputy Traffic Managers, it was Milne who was promoted to Traffic Manager, a senior administrative post. Milne, who had for so many years trailed in Brewitt's shadow, was now Brewitt's superior.[46] Upsetting the status quo was an opportunity the FBC afforded and Brewitt's upward mobility in the railway ranks would recommence soon after.

Premature retirements

During the two years that the British Empire was at war, but war with Japan had not broken out and Burma was almost as far from the fighting as a colony could be, the BR saw a spate of applications for premature retirement. For some men, the actual desire for release from service had been building up for several years. Burma's separation from India had trapped at least some white members into permanent employment in the BR when they had previously had no intention of remaining in Burma for the remainder of their careers. A small number of railwaymen, for example, had been seconded to the BR from IR service. Upon separation, they found their applications for redeployment to the subcontinent rejected.[47]

The war with Germany was both an opportunity and an additional hurdle. On the one hand, the mobilization of manpower and increased military railway traffic in Britain and a number of colonies and fronts created employment opportunities for railwaymen that had really been very difficult to come by since the outbreak of the World Trade Depression in 1929. Moreover, finding employment in Britain or elsewhere during wartime was considered a special circumstance. As a result, if Rangoon was happy to release him to find service elsewhere, the departing railwayman woud not have to pay the dreaded six-month wage penalty for a failure to give proper notice to the BR.[48]

On the other hand, the war meant that it was not just civil employment commitments that kept them in Burma, but military ones as well. Many members of the BR were members of the BR Battalion, with the Burma Army, which was a reservist formation that had been mobilized once war had broken out. This meant that men such as W. H. Durrell in addition to securing permission from the Burma Office had to also secure permission from the Burma Defence Department, although release in this case, if the permission of the governor and the Burma Office had already been secured for release from the railways, was merely another time-consuming, paper-generating hurdle. In Durrell's case, he had accepted a position as a civilian garrison engineer and hence he could accept a commission in the Royal Engineers, which would eventually take him out, but not until the war was over.[49] In A. F. Simpson's case, as a second lieutenant of the BR Battalion, Burma Auxiliary Force, he merely needed the concurrence of the Defence Department.[50] Only the onset of the war stopped the bleeding by the BR of European railway officers and staff.

Chapter conclusion

The proposed application of metropolitan military railway Transportation Technique to overseas operations did not consider the political implications of railway operations in a British colony as a factor that might hinder railway operations. The colonial railway environment was very different than in Britain because of colonial employment politics that saw white railwaymen bond tightly together despite apparent personal dislike for one another to preserve their employment privileges in a changing colonial political context. Thus, this seemingly close-knit group of railwaymen could come together and unite very powerfully for common purpose, as they did in 1942 and would do so again in 1945–7, after the war, to oppose the appointment of an outsider as CRC. Nevertheless, these railwaymen were, upon closer inspection, not so close at all. The colonial railway officers who headed Burma's colonial railways had sometimes rather iniquitously competed and fought to climb the railway hierarchy to the top, the CRC-ship being the prize. This situation will be shown in the following chapters to have provided Biddulph with an opportunity to undo the BR administration, overcome its opposition to military control or even cooperation with the military. Biddulph was then free to reform many BR officers into a militarized formation that would work in support of his military Transportation Technique, as will be discussed in Chapter 4.

3

The failure of civilian management of a colonial railway in wartime

Prior to 1942, Burma had not received the same military Transportation help as had India, but not for want of trying. When Dorman-Smith first arrived he indeed asked the British Government to lend the Government of Burma (hereafter GOB) an expert in transport.[1] But in 1941, London did not view Burma as being under any significant potential threat, protected by mountains to the east, the fortress of Singapore guarding the sea lanes to the southeast, and India on the west. British forces were strung out elsewhere across Europe, Africa, the Middle East and Asia and could not be spared for a colony that was not deemed a 'very high level of priority'.[2]

There was a knowledge gap regarding military Transportation organization and procedures among most of those involved in railway operations in the colony for some time. The BR administration was unable to cope on its own with the problems that the Japanese irruption introduced. Burma would thus depend very heavily on the arrival in early 1942 of imperial officers who could take over transport at the top, forge a Transportation organization and introduce the procedures necessary for the railways and other civil transport agencies to support the army while it fended off the Japanese invasion. These men had experience in Europe and perhaps some experience in other colonies, such as India and Iraq, but not in Burma itself. It would depend upon Brig. (acting) Francis John Biddulph, who had been Deputy Director of Transportation in Iraq, to introduce new military Transportation Technique to Burma's transport system in the process of an ongoing campaign. Biddulph's continued efforts to impress the importance of proper (imperial) military Transportation Technique, including the militarization of the railways, did not change the ultimate result of the campaign, but it did prolong the defence. Nevertheless, as Biddulph would later complain, the slow pace of this adaptation meant that the biggest obstacle military Transportation faced in the Burma Campaign was the people it was intended to help.

This chapter examines Biddulph's arrival and his effort to reorganize Burma's railway operations according to metropolitan military Transportation Technique. The introduction was slowed, however, because of an entrenched civilian railway administration that resisted change and claimed operational competence because of fears of outside interference in a changing colonial political context. This resistance persisted despite its inability to deal effectively with serious manpower shortages and demands for Forward Support, and as performance declined, tensions emerged between military and civilian users of the railways that could not be resolved.

Railway labour problems in Rangoon

Lacking the overseas commitment that India had, Burma's colonial government had to undertake the coordination of transport in the colony on its own until December 1941. Military guidance regarding the management of the BR had already been discussed in the War Office (WO) on the eve of the Japanese attacks after reports that 'important transportation problems affecting Imperial Defence have arisen in Burma'. The day prior to the outbreak of the war with Japan, the WO had communicated to the General Officer Commanding (GOC), Malaya and then repeated to his counterpart in Burma that the BR was having difficulty in handling the increased traffic and facilities. This was both in regard to the supplies going to China and meeting the defence requirements of the colony. The WO thus recommended that the Assistant Director of Transportation at the Malayan Command headquarters (HQ) visit Burma and make his own assessment. As the war broke out hours later, this visit was not made, delaying any further discussion on what to do about the BR for a few weeks.[3] Finally, in late December, a serious evaluation was finally undertaken. Raibert M. MacDougall (Indian Civil Service) was Counsellor to the Governor of Burma (1941 until 1947). As Chair of Burma's Transport Co-ordination Board, he coordinated transport in the colony and did so without help from Britain until December 1941.[4] The Transport Co-ordination Board was tasked with avoiding problems such as congestion in meeting both civil and military demands on Burma's transport infrastructure. It consisted of representatives of the different transport agencies and a Government member, the Chairman of the Port Commissioners, the Chief

Railway Commissioner (CRC), the Chief Engineer of the Roads Division of the Public Works Department and the General Manager of the Irrawaddy Flotilla Company (IFC), but it did not include a representative of the Burma Army HQ.[5] Defence in the colony and thus military planning regarding the railway was, in other words, trapped in the policy of the local administration rather than of the empire. Thus, it was *colonial* rather than an *imperial* policy that slowed the BR's adaptation to the impending demands of war up to the point that war with Japan broke out.

None of these arrangements were prepared for the scale of problems with which the transport agencies, the railways in particular, would be faced with the outbreak of the war. The biggest challenge to railway operations in Rangoon, Burma's main port and the hub of railway operations, was how to retain a sufficient labour force. Most of the railway's Indian and Burmese menial staff were paid too little to make the risks of remaining in a war zone an attractive option. Although an overland Japanese invasion (for a sea invasion was unlikely before the fall of Singapore) would take some time to reach Rangoon, the Japanese began air attacks on Burma about two weeks after war broke out. On 23 December, the Imperial Japanese Army Air Force raided Rangoon. It was a virtual abattoir, 2,000 people being killed outright, another 2,500 people wounded and, of the latter, another 750 died of their injuries.[6] This attack was followed two days later by another air raid on the colonial capital on Christmas Day.[7] There ensued a mass exodus of people, especially among the Indian and Burmese communities, out of the city. Two new Burmese terms would gain currency to refer to the people who fled Rangoon now and later. Those who fled now were *bom-pye* (air raid refugees) although the more general term that came to be used for evacuees as the Japanese advanced was *sit-pye* (war refugees).[8] This first wave of *bom-pyes* also included essential workers.[9] Wing Commander H. L. Thompson writes that 'seen by our pilots in the air, the trains and lorries running north from Rangoon resembled moving twigs on which bees had swarmed'.[10]

The most serious consequences of this migration were the reduction in available manual labour for the railways, the cutting off of the supply of food to the railways and other government workers, congestion in the Rangoon railway yards and a situation ripe for looting. On 24 December, Rangoon Station was generally out of commission because of the labour flight. The platform was strewn with bananas and vegetables where they had been dropped. Waiting military passengers went into the railway yard, hand shunted the carriages and put together a train, even finding an engine that was now stoked up, backed up and then hitched to the rake of carriages. However, the train could not budge because the drivers had fled as well. Finally, an Anglo-Indian driver volunteered to take the train north.[11]

The flight of Indians from Rangoon, in particular, created a labour crisis for the railways that it never fully recovered from during the remainder of the FBC. The railways, despite efforts to use alternative fuels, were still dependent mainly on coal imported by ship from abroad. When the air raids began and Indian labour fled, coal could not be brought in. Coincidentally, there had been a coal shortage for some time and appeals had gone out, but coal ships did eventually respond in droves only to find that there was no labour to unload them.[12] At the Rangoon Station too, there were no longer any Indian coal coolies to shovel coal into the coal wagons. As a result, engine crews, already heavily burdened with their own responsibilities, had to shovel their own coal as well.[13] Some measures were undertaken to lighten the load on engine crews and other railway staff. To make up for coolie labour, for example, Rangoon authorities attempted to use jail labour but this did not work out because, it was reported, the convicts were unused to this 'heavy and dirty work'.[14] The colony had no means to maintain railway labour, or labour for any other public work, along civilian lines. Another source of labour was clearly needed.

One of the criticisms of Dorman-Smith by American civil and military authorities in Burma and China at the time was that the Governor did nothing to stop the flight of labour from Rangoon. Colonial authorities countered, however, that there was nothing they could do. E. C. V. Foucar, whose 1943 report was generally an apologia for the colonial government's role in the defence, claimed that 'there was no effective method of preventing' the exodus.[15] Nevertheless, British military authorities complained that the Rangoon Council voted to give the Governor the authority to conscript labour but that he failed to exercise this authority.[16] Dorman-Smith appears to have sought to avoid disturbing the Indian and Burmese population, and thus refused to take this step.[17] In the middle of January 1942, a rumour circulated that Rangoon would be submitted to two weeks of continuous bombing, and a second exodus ensued which no remedy succeeded in curbing.[18] On 9 February, Tokyo Radio claimed that Rangoon would be destroyed the following day. With no official *dementi* from the government, Indians, this time mainly *dhobies* and sweepers, chose to believe it, leading to a third major flight of labour out of Rangoon.[19]

A civil railway administration at war

As a statutory corporation, the BR Board was responsible to a Burman minister regarding railway policy. However, regarding defence, the Governor of Burma had overriding powers.[20] Despite this responsibility and despite general war precautions begun with the arrival of Dorman-Smith in March 1941 as Burma's new governor, these efforts did not involve protection of the railway. Considerations of Burma's railway operations in the case of the possible outbreak of war did not envisage its operation on a conventional war front. Prior to the outbreak of hostilities in Southeast Asia and the Pacific, which came on 8 December 1941 in Burma (7 December in the United States, in another time zone), no plan had been put in place to militarize the BR.[21] Burma inherited Burma-raised units that formed its own army, separate from the Indian Army, when it separated from India in 1937. This included what became the BR Battalion, Burma Auxiliary Force (BAF). This battalion was a security force for the railways, made up of railway employees, rather than a Railway Operating Company (ROC), which had important implications for the BR when war with the Japanese broke out.

The issue of militarizing BR staff was raised in the second week after the Japanese attacks on Pearl Harbour, Hong Kong and Southeast Asia. A scheme towards this end was proposed to the Burma Army HQ in the middle of December. The scheme called for railway staff to be given weapons and military uniforms and subjected to military drill. The militarization proposal was rejected in part because, it was decided, there was insufficient time to explain the details of the militarization scheme to the railway staff. Even after the FBC came to a close and different government and military authorities compared notes, it became clear that at least some heads of civilian transport agencies were unclear about what militarization would have meant, what the structure would have been and what militarization would have offered.[22] The proposed direction of the militarized BR staff, if this step were undertaken, was also not well understood at the time or after. Some believed that this would a Directorship of Railways. In actuality, a Directorship of Transportation, covering more than the railways, was proposed.[23] Rowland had been Dorman-Smith's immediate choice for such an appointment. Nevertheless, there was some reluctance to return control to Rowland under these terms. One argument made against his appointment as CRC was that it really should go to someone with actual military experience. Another was Rowland's set of conditions for accepting the post. He required direct access to both the governor and the GOC, with his own military rank.[24] However, if he was to be given a military title 'on a par' with his civilian duties, he would have to be made a major general and this was felt to be impossible. Finally, the Burma Army HQ had its own cavil, querying privately and somewhat cryptically 'whether such a positive personality would fit in with the military set-up.'[25]

Nevertheless, Burma recognized that it would need military help with transport and on 12 December 1941 the Burma Army HQ asked India to send someone to advise on transport. This resulted in Manton's Directorate of Transportation despatching a party of officers under the Deputy Director of Transportation at GHQ India, R. Gardiner, to Burma to assess the situation. These officers interviewed high-level BR staff, including Milne, regarding what kind of procedures should be adopted to recruit staff for railway units for the colony. Moreover, Gardiner not only was a regular Royal Engineer's officer but also had years of first-hand experience with the BR and had been Government Inspector of the Railways in Burma until September 1939.[26]

W. H. Chance was Rowland's successor and the incumbent CRC of the BR since February 1941. Chance had been the leading opponent of the militarization of the BR from the outbreak of the war with Japan. But the problems that had emerged after the Japanese bombings of Rangoon had convinced him that it might be necessary to militarize the railways. By January 1942, Chance was thus not an opponent of militarization of the railways per se. Yet his adamantine opposition to the introduction of external, imperial military authorities to manage the colony's transport persisted. Instead, he sought to keep control of the railways within the colony. Admittedly, as CRC, he was best positioned to do this and thus sought a grant of special powers from the governor under the Burma Act made possible by the State of War the colony now found itself in. Chance first suggested a plan that would militarize the railways and concentrate significant powers over the railway staff into his own hands. On 11 January 1942, Chance made two suggestions for aiding railway operations. First, he suggested the embodiment of the BR Battalion, BAF. Railway staff would carry on their normal responsibilities, but their militarization would raise their morale and aid the maintenance of discipline among the railway workers. Second, Chance recommended invoking the Defence of Burma Rules for railway operations. This would allow him, as CRC, to hold over railway staff what was effectively the power of martial law.[27]

The GOB and the Burma Army rejected the proposals of Chance on various grounds. First, government ministers and the Railway Board did not think that martial law powers should be applied to a government department. Second, as the colony did not have a Director of Transportation or support staff in place, changes to the organization of civil agencies such as the railways would be disruptive. With the Japanese invasion expected anytime soon, this would be unwise. Instead, it would be better, they suggested, to complement the existing transport agencies with military formations that could provide the minimum necessary military Transportation, lines of communication (LofC) and labour support should civil agencies completely collapse.[28]

On 17 January 1942, India, in its appreciation of Burma's transport situation, recommended the creation of a Transportation Directorate, Burma. This Transportation Directorate would consist of a Director of Transportation, an Assistant Director of Transportation for Docks, an Assistant Director of Transportation for Railways and a Deputy Assistant Director of Transportation for Stores. The pre-existing Transport Co-ordination Board would be retained and placed under the chairmanship of the Director of Transportation. Left to its own resources, the colony did not have the military transport expertise necessary to undertake the proposed changes. Burma had no military ROCs. Provision had only been made at the level of the colony in Burma's case for the protection of the railway in crises. The Rangoon and Irrawaddy State Railway Volunteer Rifle Corps, renamed the Burma State Railway Volunteer Corps in 1884, underwent several more name changes until 1937, when, after separation from India, it was transferred from the Indian Auxiliary Force of the Indian Army to Burma's defence forces as the BR Battalion, BAF. Rather than a ROC, this was a railway employee recruited, volunteer rifle force, officered by high-ranking men in the civil railway administration. This force was created in the context of a colony whose main security concerns since 1885 had been internal due to several episodes of national rioting, union strikes and an insurgency in 1930–2 that nearly swept through the entire colony. It was no front-line formation. Instead, these men were intended as LofC troops and as security for the railway and stations against internal threats only. The governor, Sir Archibald Cochrane, put the BR Battalion on an active footing on 2 August 1940, but this was a nominal mobilization that had little impact on actual railway operations. Instead, it meant that the railwaymen had to join the military camp, practice on the rifle range, participate in parades, attend lectures on such topics as bomb disposal and become familiar with the Lewis Gun.[29]

Biddulph, the Director of Transportation, Iraq, had been named in mid-January as the Director of Transportation, Burma. However, he would take several weeks to arrive in the colony. In the interim, Major General Goddard in command of the LofC directed Lt. Col. F. Seymour Williams, Royal Engineers, in command of the No. 2 Docks Group, (Royal) IE in Rangoon since 12 January to perform the duties of the Director Transportation in addition to supervising dock operations in Rangoon.[30] He was assisted by the arrival at Rangoon from India on 17 January of two officers. Major A. B. Rogers of the IE, an officer who had railway stores work experience, became the new Deputy Assistant Director of Transportation. Captain N. W. G. Brown of the 7 Rajputana Rifles, who came with shipping office work experience, became the Supply Clerk, Transportation.[31]

Williams quickly held conferences with Strong, the Chairman of the Rangoon Port Commissioners; Chance, the Chief Commissioner of the Railways; and John Morton, the General Manager of the IFC and asked for reports on the position of their charges and future requirements. Of the three, the railways appeared to be 'functioning well and required no immediate assistance'.[32] Nevertheless, Williams felt that despite the positive report by Chance, bombing by the Japanese would likely require more help and thus asked for three ROCs and two railway construction and maintenance groups. Williams's hope was that the ROCs would handle train operations on three stretches of the railway line that were considered to be most likely to be bombed. The railway construction companies would then repair the bomb damage to the lines in these 'danger areas'.[33] Only one of these companies and part of another would eventually arrive.

Biddulph finally arrived in Rangoon on 27 January. Biddulph's intimate knowledge of metropolitan (British Army) organization and operating methods of military-run railways seemed to make him ideal for helping Burma adapt to the influx of various Indian transport elements and the development of coordination between them and colonial transport authorities and military formations who had experience with neither recent warfare nor with modern military logistics. Biddulph's lack of experience in Burma, or in a colony comparable to Burma, dismayed the BR men. Knowing the road and possession of 'local knowledge' was something that took years, not merely weeks or months to acquire, as discussed in Chapter 2. It is important to keep in mind, however, that during the campaign Biddulph was very clear in his correspondence to his peers and superiors and in his reports that he recognized that he lacked local, first-hand knowledge of the colony and that this would make his job especially difficult. As he explained in his report on the military transport during the campaign, he 'lacked first hand knowledge of the country, transport systems, and the personnel, civil or military, with whom he had to deal'.[34]

To overcome the local experience gap, Biddulph requested that Gardiner at GHQ India in New Delhi be sent to Burma as the Deputy Director of Transportation, but was told that Gardiner could not be spared. Soon after his arrival in Burma, Biddulph would also recommend that Rowland, already rejected once for the post, should be commissioned and made Director of Transportation and Biddulph made Rowland's deputy. Biddulph argued that because of Rowland's reputation, his possession of a knighthood and his already having been CRC, he had more status and authority than Biddulph and hence would have an easier time soliciting support and cooperation from the BR staff and possibly could have halted desertions from their ranks. Biddulph had even requested that Rowland come down from Lashio to Rangoon shortly after his arrival in Burma to consider taking the post. Although Rowland did come down to meet with Biddulph, neither the army nor Rowland would agree to this proposition.[35]

There were other factors involved that would have caused local resentment over the appointment of any outsider. As Director of Transportation, Biddulph was placed in overall charge of the railways, through the Director of Railways, whom he appointed, and in supporting the latter's decisions regarding the railway staff's performance of their daily responsibilities as well as their professional careers. Biddulph, as an outsider, apparently had no personal or professional connections with BR staff prior to his appointment and was thus in their eyes at best an accretion. He lacked the personal networks that officials in the colonial bureaucracy tended to depend on for mutual security in administrative infighting and as aid in securing promotions and healthy pensions.

Further, Biddulph was unacquainted with the in-house management culture. Isolated institutions like the BR, dominated for decades by the same management, often develop peculiar ways of doing things. In fact, every railway company or national railway developed its own railway culture and these were forces to be reckoned with. McKenna's observation that early nationalization of the railways in post-war Britain found it difficult to weld together into a singular whole 'an amalgam of tongues, an entrepôt of railway cultures' inherited from the era of private companies bears this assertion out.[36] Neither technical manuals nor experience in other railways could fully prepare a stranger for it. Any officer, whether military man or railwayman, brought in from the outside, would have found it challenging to communicate or do things in ways that would not seem foreign to the BR's perdurable operating culture and would have encouraged a feeling that the officer was either inept or peremptory and certainly that he was 'not one of us'.

Chance's railway administration attempted to meet the demands of increasing military traffic by undertaking several measures that gave an 'emergency'-style façade to traffic control. In January, the Railway Head Office and control room were set up in Kambe, six miles out of Rangoon. Brewitt was in charge of the control room, where he managed the movement of stock. As the Japanese got closer to Pegu, the work in the control room at Kambe became 'completely disorganized' and the main office was again moved to Rangoon, where an officer was on duty twenty-four hours a day. Here, Milne and others slept where they served and directed the evacuation trains and the running of special military trains.[37] However, these steps simply added more administrative confusion without offering any demonstrable efficiency to operations.

More than the civilians, the railway administration was lacking in meeting the demands of wartime. Eric Norman Goddard, since December 1941 the major general in charge of administration, Burma Army, would later admit to Biddulph as would many others that he had no knowledge of the subject of military Transportation Technique.[38] Goddard recognized that outside military help of some sort would be needed even before Biddulph was scheduled to arrive and he appealed to India for more help in Burma to deal with the increasing transport demands of the war. It requested Inland Water Transport (IWT), dock and railway units be lent, the latter to include a Railway Construction Company and a Railway Survey Section. At the time, a very limited number could be spared from India. The first two units, out of a total of only three that were eventually sent before March 1942, were dock-operating units, which arrived in Burma on 12 January 1942. Any other help, it was suggested, had to be formed out of staff already available in Burma.[39]

After he arrived at the end of January, Biddulph outlined his plans for a Transportation General Directorate; recommended that it include sub-directorates for IWT, docks and the railways; and that it should be put into place when it was possible to do so from local resources. He then undertook separate discussions with the heads of each of the transport organizations to obtain staff for the sub-directories, but no staff could be lent. Biddulph held out hopes that the withdrawal from Tenasserim would yield surplus staff from which to form his new organization.[40] Biddulph also suggested that the transport agencies be militarized immediately. As employees of civil agencies, he argued, he was unable to feed or pay railway personnel and they had to rely on local shops that were no longer open in the Rangoon area. According to S. Woodburn Kirby, writing some years later, Biddulph's recommendation was rejected because it was too late for this to be effective as the Japanese had already reached Moulmein.

Moreover, when Biddulph advised that the BR Battalion, BAF be embodied in early February, a few days after his arrival, Chance, the CRC, rejected the proposal outright.[41]

Biddulph offers a different account in his 'Narrative of the Military Railways in Burma'. According to Biddulph, Chance at first responded favourably to the proposal to embody the BR Battalion as well as to the suggestion that the BR Battalion's expected role be changed from defending the bridges and so on, to undertaking the technical working of the railway. Biddulph hoped that by stretching this battalion over the whole railway network, it could provide officers, warrant officers and non-commissioned officers who could serve as the nucleus for the formation of new military railway units in each area of the network. In areas where some leisure hours were available military training in self-defence could be provided, perhaps by those railway officers who had military experience during the First World War or military training at some point afterwards. If retreat continued, railway units could be withdrawn to the rear area where they could be given military training and then returned to the Forward Areas. The railway officers whom he consulted were in favour of Biddulph's plans and were eager to play a bigger part in the defence effort.[42]

The resistance of the Chairman of the Port Commissioners and the head of IWT killed the scheme by dragging out discussions interminably over details. After being consulted by the Chairman of the Port Commissioners, Goddard wrote to Biddulph that in the present conditions there was 'no question of the Army taking over the civil Transportation Agencies'. Goddard was also against militarizing civil transport officers so that they could organize their own directorates. As Goddard wrote, 'I have been unable to convince myself that this is really necessary.'[43] Chance then changed his mind as well and wrote to Biddulph on 7 February 1942, explaining that he had decided that the disadvantages now outweighed the advantages. As Chance pointed out, embodying the BR Battalion would not make so much of a difference to railway operations because members were a very small proportion of the 22,000-strong railway labour force. In some cases too, military pay would be higher for certain individuals than for others doing the same work but who had not been mobilized, promising to create resentment. The most important reason, in Chance's view, was that he had agreed to the embodiment of the battalion mainly to bolster staff morale but in the interim had proven they could handle difficult circumstances and were thus fully prepared for war conditions without the need to militarize.[44] Something else had also changed, at least in theory. The application of the Defence of Burma Rules to the BR now empowered Chance to force railway employees to stay at their posts. As Biddulph later pointed out in his narrative account, however, although legally Chance could have any employee who disobeyed his orders imprisoned, the machinery to apprehend offenders was non-existent and hence the power was 'inoperative in practice'.[45] Kirby has also added that the threat of imprisonment did not make up for inadequate training or discipline. Thus, railway workers continued to abscond and the railway's problems grew worse.[46]

The challenge of geography and the railways in Tenasserim

The geography of the lands around the Bay of Martaban also complicated railway performance during the remainder of Chance's administration. Initial Japanese operations took place in Tenasserim. A Japanese force entering Burma via the Kawkareik Pass threatened Moulmein and along with it the entire Ye Branch of the railway.[47] Here, the BR men had their first taste of running a railway up to the front lines, but the experience went badly. The route of the Japanese attack came as a great surprise to Burma Army planners. Tenasserim was not considered a high-priority area again because it was believed that the Japanese would come through the Shan States and not the south. As a result, little was done to improve the branch lines in this area to provide Forward Support for the Burma Army for military operations. Nevertheless, when the Japanese invasion came, the retention of these branch lines became crucial to Burma's defenders, although their strategic value would be severely compromised once Singapore fell in mid-February.

Although served by a railway, locomotives and rail stock in place in southern Tenasserim were meagre, reflecting the secondary importance of the area to the colonial economy. The railway here was merely a branch line that extended another branch line. Railway passengers and stock coming from the north were detrained and unloaded at Martaban, loaded onto a pontoon ferry run by the IFC, then brought across the estuary to a small station at Moulmein on the east bank and then entrained and loaded onto the Ye Branch. The Ye Branch, running south from Moulmein and the east bank of the Salween Estuary, then took trains as far south as it was possible to go by rail in Burma.[48] There was, however, not a great diversity of traffic. Tenasserim generally had always been of secondary importance to colonial Burma's economy, although the early settlement of Europeans in the littoral, especially of missionaries at Moulmein, made it of special interest to Rangoon for other reasons. The main load going north was salt to meet domestic consumption as Ye supplied Rangoon and the rest of Burma with the substance. Almost no loads were brought to Moulmein for export abroad. Goods did come south to meet local demand, being unloaded onto flats at Martaban, carried across the estuary and then loaded up into railway wagons on the Moulmein side. A special flat was also kept on the Martaban side of the river for use twice a year, when tides were suitable, to bring locomotives and rolling stock north from the Ye Branch to the Martaban Branch for repairs, the steamer used to tow the flat being hired, again, from the IFC. Facilities along the branch were thus second-rate, small and under-staffed stations strung along a straggling line far distant from Rangoon. The rail line on the Ye Branch was also the weakest in Burma. On all other branches, the permitted axle load was ten tonnes, but here it was only eight tonnes.[49]

The early loss of Moulmein ended BR operations on the Ye Branch, as the rest of the line to the south was already entirely in Japanese hands. All six locomotives on it were left to the Japanese in working order. There had been arrangements in place to deny these to the enemy, but on instructions from military authorities, staff abandoned the branch before denial could be carried out.[50] Such early military reversals in the campaign were an indication of the difficulties ahead for the BR and were similar to those of other railway systems providing Forward Support for any defender in fast retreat. Despite the challenges, the military needed the railwaymen to remain steadfast in providing this support, for to have the railways fail would deliver a military disaster. According to the official WO (London) version of the events, the flight of the BR men was actually the cause of the Burma Army being forced to abandon the defence of Moulmein. In this view, Burma's railwaymen proved unreliable in the face of battle and the drivers and others deserted their trains and other posts. There is little information on railway operations from the perspective of the BR men active in this particular operational area to confirm or deny these statements, but given the shock of the invasion to untrained railwaymen who did not expect an invasion from this quarter, it should not be surprising if this was indeed the case.

The Burma Army would soon discover the difficulties in defending the Martaban Branch. This branch left the main trunk line at Pegu as it moved out to the east. The branch from Pegu ran up to Martaban, a small station served by 'some rudimentary' marshalling yards,[51] on the west bank of the Salween Estuary. There were several natural bottlenecks to the extended journey, from Pegu to Martaban, the most significant being the Salween Estuary and the Sittang Bridge, the latter referred to by Hingston as 'the greatest bottleneck in southern Burma'.[52] However, there were also numerous smaller points of high vulnerability all along the Martaban Branch, including the bridge at the Bilin

River. The BR moved the first British unit deployed in the FBC, the 2nd Kings Own Yorkshire Light Infantry Regiment, from Takaw to Martaban to help reinforce the collapsing Burma Army units, on 23 January 1942.[53] From this point on, the traffic demand, soldiers moving south, and soldiers, civilians and wounded moving north overwhelmed an unprepared colonial railway service.

Many BR men in Tenasserim also performed poorly. Just as the railway system itself was threatened, so too were the lives of the railwaymen themselves by service close to and sometimes ahead of armies on the front lines. Despite the risks, few railwaymen had recent training in the operation of trains in a combat climate. Although some railwaymen in Burma had experience in the First World War on the Western Front and in the Balkans, that had been very nearly a quarter of a century earlier. Railwaymen who had been trained as railway operators now found themselves shuttling troops and wounded back and forth, operating trains under fire, fending off Japanese soldiers with rifles from embankments and inside besieged trains and even tasked with destroying their own workshops, locomotives and other equipment. Often their bravery in these initial Front Area activities gave way to fear and many fled. Given the absence of BR men on this branch that resulted, the Chief Railway Engineer of the 17th Indian Division was forced to supply military train crews made up of divisional engineers in order to effect necessary train operations in support of the army. Given the dire situation, there was a surprising degree of resistance to this. The few remaining civil railway staff and the railway administration protested because these military engineers were not actually trained in railway work and thus there were risks of accidents.[54] Nevertheless, other desperate soldiers took matters into their own hands. Among those who waded ashore on 31 January was Capt. D. G. Lacy Scott who had an understanding of how to operate a steam engine and was encouraged by finding a locomotive at the head of a long troop train. Confident he could drive the engine himself, his group approached the engine only to find the civilian train driver and fireman hiding from the Japanese mortar fire from the Moulmein side of the estuary by taking refuge underneath the locomotive's boiler. Persuaded to drive the train, they got onto the footplate, 'set the cut-off to full forward, opened, closed and re-opened the regulator' and the train with Scott's party proceeded on to Kyaikto.[55] Such actions had to be undertaken because of the failure of the BR to man the engines and get the trains running at Martaban.

Formations were now digging in to await the resumption of the Japanese advance, but within a few days, a complete withdrawal of Burma Army troops and railway assets from the littoral began. As was being experienced at

Martaban, railwaymen and railway equipment were part of the forces moving back, sometimes in an organized way, but increasingly commonly were closer to a disorganized mass of men desperately trying to escape and using any form of transport to get out. The Japanese began bombing railway stations on the branch as well, beginning on 2 February with an air raid on Yinnyein Railway Station and a nearby quarry (the Martaban Branch was one of two lines in the colony that served the quarries supplying the BR with ballast, the other being the Shwenyaung Branch). When Martaban was captured by the Japanese, the District Traffic Superintendent at Pegu on the other end of the Martaban Branch, U San Paw, 'went sick'.[56] Generally, BR men were failing to provide for timely Forward Support for the Burma Army all along the line. Further delays in getting help to support the Burma Army in Tenasserim, however, were imposed by the same problems in Rangoon that hampered railway activity in the colony generally.[57]

February saw the continuing military retreat that ended with the loss of the entire Martaban Branch. On some trains the mood was so low that fear of desertion led to the issuance of orders that senior non-commissioned officer in each carriage on the trains post sentries at each end and not permit anyone to leave the carriage whenever the train made a stop along its journey.[58] When train activity of any kind had ceased on the Tenasserim side of the Sittang Bridge, Royal Engineers converted the railway bridge over the Sittang River to the use of motor vehicles as elements of 17th Division that needed to withdraw over the bridge were in part carried on lorries. All locomotives on the Martaban Branch had been successfully taken off of the line and across the Sittang Railway Bridge, Burma's second-largest bridge, before it was blown on 23 February.[59]

The loss of command of the sea for the railway

The news of the fall of Singapore the previous week, on 16 February, affected the railways in a number of ways. It was first of all a blow to the morale of the defenders as a whole.[60] British propaganda had built up the impregnability of the fortress to such a degree that its surrender caused a wave of demoralization to spread over the breadth of Burma, disheartening the military, civil government and civilian population alike.[61] It also exposed Burma to a possible sea invasion. So long as Singapore held, the rollback of imperial forces up the Tenasserim Littoral was not fatal to Burma's defensive capabilities. The lowland coastal strip on which these forces fought was narrow and still on the right side of the Sittang River, allowing even heavily demoralized and reduced forces to put up

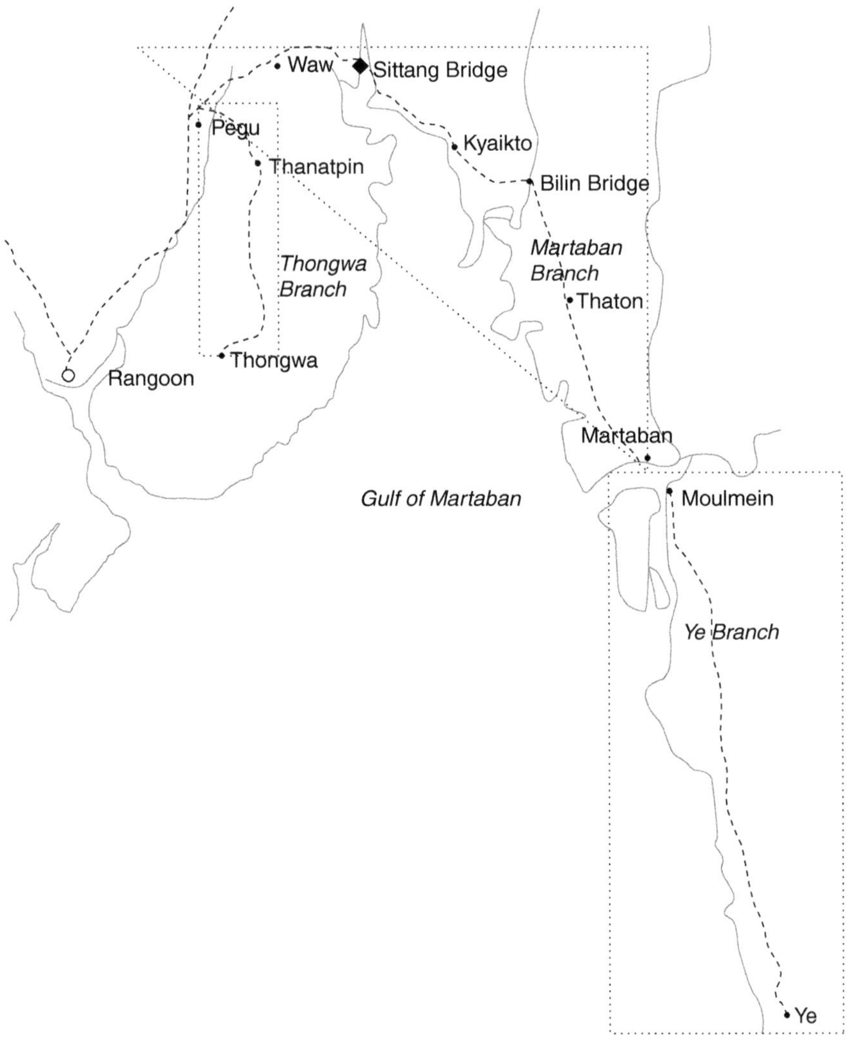

Map 3.1 The Burma Railways and the war in the Southeast, January–March 1942.

meaningful resistance. However, when this rollback coincided with the fall of Singapore, the Tenasserim Littoral ceased being simply a land route north and now became a potential base from which the Japanese could ferry troops across the Salween estuary and Bay of Martaban by small boats and land on the coasts between Rangoon and Pegu. They could then either occupy the area and hinder river communications to Rangoon or destroy the Burmah Oil Refineries at Syriam that handled the oil piped down from Yenangyaung on the Irrawaddy River north of Prome.[62]

Defence units formed together as 'Pegu Force' were now tasked with protecting the eighty-mile strip of coastline, defending the southern approach to Pegu where it linked up with 17th Division and guarding the coastline on the banks of the Rangoon River itself near Syriam.[63] This necessitated mobilizing a railway branch of even lesser importance than those thus far discussed in Tenasserim and expanding the area of necessary Forward Support. The main trunk line between Rangoon and Pegu ran just north of the Pegu River and then along the northern edge of wet, paddy lands. These paddy lands were effectively impassable for motor transport. As a result, this railway branch and that of the Martaban Branch between Pegu and the Sittang Bridge essentially served as a boundary, dividing the paddy land to the north and west that was passable to motor transport and that to the south and east which was not. This area of land along the coast ran southwest to form the east bank of the Rangoon River, where it hosted Syriam on its far northwestern tip.[64] The Thongwa Branch ran south from Pegu through the centre of these paddy lands to Thongwa Station.

One of the worst aspects of the BR was its domination by long, overstretched branches of little importance that wasted men, locomotives and rolling stock. The Thongwa Branch was one of the worst examples of this. The branch, from Pegu to Thongwa, covered ninety-six miles of railway, while a road between the same two terminuses totalled only twenty-six miles.[65] Biddulph saw the branch as merely a 'dead-end' and of 'no military importance', although he assessed the area as it appeared in survey maps and not with direct, local knowledge of the area, which would have indicated why such a branch would retain some value in defending the approaches to Rangoon.[66] Moreover, local railwaymen were aware that the railway provided the only telephonic communications in the area and this was necessary for communications between Pegu Force HQ and police platoons. In this part of the colony wireless sets were unavailable. The country was too flat to allow for communication by lamps or heliograph. Motor patrols visited the different posts each day, but as mentioned, the area, aside from the road that went down the centre of – and not around – the peninsula, was considered un-motorable. As the regimental history explains, bullock carts had made deep ruts on already rough tracks and bunds separating paddy fields had to be dug away just to provide sufficient clearance, forcing the motor patrols that dared to risk the journey to travel at about five miles an hour to cover distances of twenty or more miles. The Pegu Force had to rely instead on the railway's telephone line and attached mounted orderlies to each police platoon to carry messages back and forth between the platoon and the closest railway station.[67]

The defence of this area also led to the creation of Burma's first armoured trains of the FBC. Field Marshall Archibald Wavell wired Hartley on 23 February suggesting that divisional defence was not the best method to hold the Pegu area now. Instead, it was better to divide the troops into 'widely separated mobile brigade groups or even smaller columns' that would use the Thongwa Branch as a base. This would allow them to converge rapidly whenever the enemy appeared between the railway and the river. As Wavell argued, 'by such methods wide areas can be covered offensively and enemy infiltration tactics defeated by immediate attack from several directions'.[68] These tactics, it was suggested, would also benefit from the application of improvised armoured trains.[69] What constituted an armoured train was unclear, and when the West Yorkshire Regiment, which was now ordered to proceed from Yitkan to the terminus of the branch at Thongwa thirty miles away, put together such a train it was only considered to be 'armoured' because it included an open wagon in front and on the rear end, each mounted with an anti-aircraft (AA) gun. Ultimately, the regiment did not linger long at Thongwa because the railway branch here, like the Ye and Martaban branches, was vulnerable to being clipped off from the main network by enemy action. As the Pegu River to the north cut off the peninsula from the rest of Burma and Pegu to the northeast was in the path of an impending Japanese attack, the only way out for the regiment if the situation soured was a bridge at Thanatpin. The regiment thus kept two flats at Thanatpin ready for conveyance over the railway bridge when need arose.[70]

Biddulph's position was that there were clear benefits for railway operations to be gained by abandoning the Martaban and Thongwa Branches. The rollbacks on the Ye, then the Martaban and finally the Thongwa Branches cut down railway mileage and kept railway assets continually concentrating up in a backwards flow. These surplus resources would help to replace losses through attrition for the first weeks of the campaign. Retreat, Biddulph would later argue, became a crucial enabler in the railwaymen's success in keeping the railways working. We will return to this argument again in Chapter 8, but it should be mentioned at this point that had the defence stabilized or, worse, had a counteroffensive against the Japanese succeeded and the latter were rolled back, Biddulph was convinced that the railways would have been unable to support the army. The sacrifice of space, in the case of the railways, mileage of permanent way, benefitted continued railway operations.[71]

Nevertheless, both the Martaban Branch and the Thongwa Branch intersected with the Rangoon-Mandalay Trunk Line that passed by Pegu. The decision to hold the line against the Japanese on the Sittang River and not forward of it would thus necessarily have very serious implications for the railways and thus for both the supply route to China and for the evacuation plans for Rangoon. As Gen. Thomas J. Hutton, General Office Commanding, Burma Command, had already reported to the American-British-Dutch-Australia Command (ABDACOM) on 18 February, holding the Sittang, so close to the arterial Rangoon-Mandalay Trunk Line, threatened to 'interrupt or seriously interfere' with the use of the railway through Pegu.[72] Once the 17th Indian Division had withdrawn behind the Sittang Line, the front lines were very nearly pulled up to the Rangoon-Mandalay Trunk Line, which was now vulnerable to being cut by the Japanese. The loss of the Rangoon-Mandalay Trunk Line would be very damaging to railway operations, as Rangoon was the main hub of railway operations, railway repair facilities, coal stocks and railway administration. Worse, if such a cut took place, Rangoon would be isolated from the rest of Burma, dooming it and the colonial government to capture. On 20 February, Dorman-Smith ordered the first stage of the three phases of the evacuation of Rangoon to begin. The initial expectation was that Rangoon would be completely evacuated by 2 March with an order to completely give up on the city depending on the course of events on the front lines.[73]

Biddulph held out little hope that the railways, which had thus far failed miserably, would fare any better as the Japanese approached and demands on the railway increased. For the BR men, the Japanese bombings of the railway at Toungoo and Swa on 22 and 23 February were additional, tremulous experiences and many now deserted.[74] By this time, the railway 'had already almost ceased to function owing to defections amongst the subordinate staff'.[75] Although the Japanese had been crossing the main trunk line already, the inability to close the gap between Toungoo and Pegu effectively permanently bifurcated the BR into a southern half which quickly wilted and then died and a northern half that would more gradually shed its mileage over the following two months. Burma's fate was being decided by actions taking place at the important communications centre of Pegu. Here, three routes ran through the town. Maintaining the railways as a single transport system would have required holding onto this town. The Rangoon-Mandalay Trunk Line ran from the southwest up the left bank of the river, dividing northwest of the town into the Martaban Branch that crossed the railway bridge on its path to the southeast and the trunk line that kept moving northwards alongside the Pegu River. The road to Moulmein joined the road from Mandalay in the eastern quarter of the town, the latter continuing on through, crossing a road bridge over the Pegu River and then crossed the railway trunk line before also veering to the southwest to Mingaladon in the northern part of Rangoon. The Pegu River ran roughly north to south hugged by the town centre.[76]

The loss of Tenasserim and the Thongwa areas was mainly significant to railway operations in the FBC not because of these areas' intrinsic value regarding resources, material or human, but because of their strategic value. They were geographic buffers that shielded the main trunk line for the railways, the main colonial port, as well as population and political centre and the approaches to the main river artery, the Irrawaddy. Their loss culminated in the Japanese taking of Pegu, the linchpin to the defence of Burma's southeast. What made these areas difficult to defend was the local geography. Wet, marshy alluvial soils intersected by a myriad of tidal creeks, very similar to the coastal lands around Burma as a whole, made the Thongwa and Tenasserim areas difficult to defend. A lack of colonial investment in these areas, because of their limited potential economic value when compared to other areas of the delta, meant that infrastructure here was limited and where it existed was often ramshackle. Yet, the railway provided almost the only means of moving the Burma Army, as without roads these areas were very unsuitable for motor transport. This was the main reason why the 7th Hussars and 2nd Tank Regiment had to wait patiently on the more solid soils and flat plains of Pegu, on the edge of the wetlands, for the Japanese to attack rather than push forward. By that time, the Japanese thrusts were too close to the railway trunk line to save the day. The loss of control of the sea reinforced the railway's importance in what remained of the fighting in this part of the colony. When the Japanese took Singapore and entered the Indian Ocean, whatever military assets the Burma Army had remaining in area, whether forced into retreat or not, would have been entirely dependent upon the railway. They would have hugged, as the Burma Army actually did in their retreat, the railway line all the way north, whether moving in a railway coach or on foot.

Chapter conclusion

The direction of the Japanese attack, up through Tenasserim, was not necessarily a problem for railway operations. In fact, the abandonment of the railways in the southeast deprived Burma of a lengthy railway line that before the Japanese construction of an overland link with Thailand in 1943 was of little strategic value. It also kept railway assets moving north and then west, helping to provide surplus engine, stock and personnel in the early months of the FBC. However, the cutting of the trunk line from Rangoon to Mandalay was a serious blow that threatened railway operations altogether and was part of a larger crisis facing the railways at the end of February 1942. This crisis saw the civilian railway administration's declining performance and failure in providing Forward Support as the Japanese moved north through Tenasserim towards the Irrawaddy Valley.

The railways, under greater threat than other transport agencies before the fall of Rangoon, were the military's biggest problem among the three transport agencies in the FBC in part because they were potentially the most valuable. The railways' capacity for moving people and supplies was vastly more powerful than the road system in the colony, and by comparison to the IFC's fleet of steamers whose operational areas were limited of course to the rivers and coasts, the reach of the railways was more extensive within the boundaries of the colony. Moreover, the area of Burma most immediately threatened by the Japanese, the eastern part of the lowlands running up to Mandalay, was in particular best suited by the railways. Thus, while losing Pegu meant severing any connection between Rangoon and Mandalay altogether, whether by railway, motor vehicle, bullock cart or boat, its most damaging prospect was the blow it would deal to railway operations in the colony.

As this chapter has shown, there were very clear reasons to militarize the railways and perhaps other transport agencies after Burma was invaded. Nevertheless, entrenched civilian administrators at the head of each major transport agency in Burma slowed Biddulph's efforts to introduce metropolitan military Transportation Technique to Burma from late January. They resisted military supervision and indeed there was conflict between civilian and military users over the use of the network. This resistance persisted despite the inability of the railways to deal effectively with challenges and obstacles related to the wartime situation, especially labour shortages and an inability to meet the demands by the military for Forward Support. This phase of declining performance and failure in railway operations saw the loss of the southeastern portion of the railway network and set the stage for Biddulph's seizure of control of the railways, as discussed in the following chapter.

4

Militarization and the Rangoon to Prome Branch

Reliance on reports made by civilian heads of transport in Burma meant that the serious problems that Biddulph found in Rangoon were a surprise. The Transportation Directorate, India, had been assured by Chance, through Williams, that the railways were doing well, could handle the ongoing challenges of the war and needed no outside interference or management. Biddulph outlined the actual conditions found in Rangoon when he arrived at the end of January in his 'Narrative of the Military Railways in Burma', written sometime later. Martaban, he explained, was on the verge of collapse at the time. The Burma Army was withdrawing from Tenasserim. However, if desertion occurred on even the most moderate scale, the depleted levels of military personnel could not fill the gaps. He had complaints about the military command in Burma and awareness of how transport should be directed, but that problem could wait, because the biggest and immediate problem was the reliability of staff.[1]

Biddulph attempted to resolve this most pressing and immediate of problems with a soft approach to dealing with colonial authorities and then a more heavy-handed one, first by attempting to work with the civil railway authorities and then by removing the civil railway administration from control of the railways in Lower Burma. This changeover in approach took about a month, the timing determined by the rollback of the Burma Army, the collapse of the civilian railways and the imminent Japanese sweep across the Irrawaddy Delta. Simultaneously, Biddulph sought to bolster the BR officers and staff with contingents of Indian and Royal Engineers but was only partially successful, although the men proved to be invaluable.

This chapter examines the reasons why Biddulph was afforded an opportunity to fully introduce Longmoor Transportation Technique and Transportation organization to Burma, over two full months after the war with Japan had broken out and just as the Japanese were about to move into the Irrawaddy Valley. As

Biddulph's efforts showed, it was by no means a given that innovation would be accepted by those in need, even when it was very clear that the current ways of doing things were failing. Change threatened to upset the dominance of local, colonial elites over the railways and it was the resulting tensions and resistance that delayed the introduction of military Transportation Technique until it was perhaps too late to save the situation.

The arrival of the IE

In early March, Manton, writing from New Delhi, remarked to Biddulph that he had at first reacted with surprise to the news that Biddulph was in Rangoon. Rangoon was not the problem, Manton suggested, for major Transportation difficulties should have been in Upper Burma.[2] But engineering support provided by India to maintain Rangoon was both sparse and slow to arrive. The Headquarters (HQ) No. 3 Railway Construction & Maintenance Group (RCMG), (Royal) IE, under the command of Prendergast, arrived at Rangoon from Madras on 6 February after a five-day journey at sea.[3] All of the men in the unit were volunteers sent as an emergency measure and deployed only when it was believed that the Japanese were about to invade Rangoon by sea, to aid in the evacuation of Rangoon by railway.[4] This group consisted of civilian draftsmen, tradesmen and clerks who had mainly been recruited into the group in November and December 1941. In that time, although they could acquire very little military training before being sent to Burma, they would be applauded after their service in Burma for the coolness and bravery they evinced in the most difficult of circumstances.[5]

Organizationally, the No. 3 RCMG was attached to the Advanced Army HQ then at Rangoon, working in connection with the Director of Transportation, Burma, with Prendergast acting as Liaison Officer with the BR. This relationship and possibly an earlier working relationship in Iraq led to Biddulph appointing Prendergast as the Assistant Director of Railways. This group was put to two main tasks. First, it populated the railway system where needed to both supplement the dwindling railway staff ranks and maintain necessary functions and served as an all-round morale stop-gap to prevent further absconding. Between 6 and 14 February, its work consisted of preparing siding schemes. On 15 February, the group's personnel all volunteered to help work the railways, taking up posts as cabin-men, yard assistants and telephone operators. For the next week, they worked very long hours trying to keep the railways running in the midst of evacuation and a deteriorating situation. When necessary, they made ad hoc arrangements. On one occasion, when faced with a complete absence of labour in the main railway yard in Rangoon, Prendergast went to Advanced Army HQ and obtained packets of one-rupee notes that he carried to a crowd of Indian refugees outside the Rangoon Station. He adduced the reasons they should help such as the fact that if they did nothing, no more trains would be running and they would remain stranded in the port city. With pay and pleas, Prendergast persuaded 200 men to work on the railways in eight labour gangs of twenty men each. They cleared the piles of junk left abandoned on the platform by evacuees, cleared out wagons of any non-military items, loaded coal and did other miscellaneous work. The group's orderlies also went to work cooking and taking food to the railwaymen.[6]

No. 3 RCMG's other main task was to prepare on Biddulph's behalf for the militarization of the BR once the opportunity came. Chance's reluctance not only to relax control of the railways but, in fact, to also increase his powers in the crisis served as an obstacle which prevented Biddulph from direct management of the railwaymen. The group's records document it being put to work in developing various schemes to organize BR personnel into railway military units during this period. Unfortunately, as with so many other valuable materials destroyed in this campaign, the associated documents were lost during the final evacuation.[7]

Different events are viewed as central to the experience of various participants in Burma's defence against the Japanese invasion in 1942, aside from the trek to India following the FBC, which most parties accept as a very significant and traumatic episode. For the Burma Army, the destruction of the Sittang Bridge and the encirclement at Yenangyaung are of greater significance than the evacuation of Rangoon. For the American Volunteer Group and the Royal Air Force (RAF), the key watershed event was the two-day bombings of Magwe Aerodrome in March. For the civil administration, the evacuation of Rangoon and perhaps the bombing of Myitkyina on 6 May seem to be key. By contrast, the BR men's main reference events for their experience in the FBC were the stages of militarization, beginning with partial militarization on 25 February, full militarization on 15 March and the revocation of military control over the railways on 15 April. It would be an exaggeration to say that no other events, such as the evacuation of Rangoon and the destruction on Mandalay, structured their experience more than did militarization. However, while the evacuation of Rangoon and the 3 April 1942 bombing of the railway yards at Mandalay were fantastically challenging events for railwaymen on the spot, militarization was an experience that helped to weave together the invasion experience that informed BR men's collective memories more generally. The railwaymen committed almost no comment as to why various railway stations should have been exposed to bombing or why Rangoon should have to be evacuated. However, they did direct their attention to the varying degrees of militarization, and ultimately, militarization received the lion's share of criticism as an obstacle to railway operations, even more so than the Japanese invaders.

Biddulph's assessment was that militarization had been opposed in Burma for many of the same reasons that gave him difficulties when introducing a Transportation Directorate in other overseas postings. Indeed, he saw the Burma experience as a 'repeat of India' during his time there early in 1941. Local colonial officials had suggested that Burma 'had nothing to do with India' and that his plans were nothing more than 'part of a despicable and grandiose plan to make Transportation predominant' in the colony. Biddulph explained that he had tried to admit ignorance of the situation in Burma and thus solicit help and cooperative advice from local railway officers but was instead greeted with 'distrust and suspicion' and experienced the 'same frustration' he had experienced in India.[8]

Those most vociferously negative about militarization were the railway authorities negatively impacted by the personnel changes. As we have seen in the previous chapter, although the railway administration under Chance at first resisted militarization, it seriously considered self-militarizing under Chance's leadership when pressures to militarize grew. Foucar and Dorman-Smith who relied most on the testimony of these men shared in the negative appraisals of Biddulph's efforts, however much they may have sympathized with the challenges he faced and appreciated the conditions under which he worked. Dorman-Smith felt that the task of running all of the transport in Burma in a wartime situation had been too big of a job for Biddulph. He believed Rowland should have been put in charge instead. Dorman-Smith also hinted at the friction involved in colonial civil authorities and military authorities working together, claiming 'instead of having some dynamic personality [Rowland] in charge, we had to put up with a delightful colonel [Biddulph]'. Dorman-Smith also complained that the QMG, Goddard, had not been 'pleased to be working with civil authorities'.[9] Acidulous criticism of Biddulph, although often voiced in civilian–military terms, actually owed much more to the fact that outside imperial authorities were intervening in local, colonial railway management. Ironically, although Biddulph became in the end a scapegoat for the putative failure of the railways during the campaign, it was clear that the railway as *managed* by the local railway authorities based on colonial administrative experience without Longmoor experience had indeed failed.

By contrast to the disappointing performance of the colonial railway management, the militarized railway's performance was outstanding. Touched by the afflatus of necessity, Biddulph's intervention was timely and his selection of men from among existing railway personnel was wise. These actions and the organization of these men into a military Railway Operating Company (ROC) with the aid of some Indian military railwaymen arriving in Burma from Jullundur together saved the day. These cumulative developments, wedding together local railway experience and imperial military railway organization and Transportation Technique, got the railway back up on its feet. The railways would support the defenders and evacuees of Burma for seven more weeks, under conditions far worse than those which convinced Chance to call it quits, before handing the railways back to civilian management a few weeks short of its final collapse. By any fair measure, the militarization of the railways was a resounding success and its survival until the second week of May (when the Japanese gained complete control of the parts of Burma serviced by the railways) was equally a magnificent achievement. The present chapter examines why militarization followed the course that it did, from partial to full militarization, and why, in the end, it was returned to civilian hands.

Partial militarization

Biddulph had considered and recommended militarization since his arrival in Burma and as the railways and other transport agencies had trouble maintaining operations as February wore on, Biddulph continued to argue the militarization case. When Goddard went on a one-week tour of the Front Area he came back with a changed attitude and now agreed that militarization was necessary. As Biddulph found that the transport agencies were now, ironically, even less favourable to militarization than they had been a few weeks earlier, he put forward militarization proposals in draft form directly to the Defence Department for approval without the consent of the heads of the civilian transport agencies.[10]

Biddulph's initial plans for militarization were outlined in a memorandum submitted to Burma Army HQ on 13 February 1942. He outlined a plan by which separate military directorates would be created for railways, IWT, and Stores and suggested that a later one might be created for roads. These directorates would operate as cells within existing transport organizations. The heads of these cells would be directors who would hold commissioned rank and would continue in their civil appointments so far as possible and for these functions only, they would remain responsible to their civilian superiors. It was considered desirable to have as many of the other personnel in each organization volunteer to enlist as possible. Military units would be formed from such personnel especially for work in operational zones. Additional transport personnel added to these organizations to meet military need would be by adding military transport units. By virtue of their military status, directorate personnel would hold powers of command over these military transport formations. The Railway Board, Burma, the Rangoon Port Trust and the IFC were invited to submit the names of candidates to serve as directors and after their selection, they would consult with Biddulph on the establishments they required.[11]

As the plan awaited formal approval, Biddulph submitted copies of the same memorandum to the heads of transport organizations on the following day, to speed up movement on the plan as a matter of urgency, along with a lengthy letter explaining why this course of action needed to be undertaken. Transport, he told them, was 'a vital factor in war', and the commander-in-chief had a responsibility to maintain the army's lifelines as much as he is responsible for its success of the battlefield. To discharge this responsibility, he has to exert a degree of military control equal to the scale of the charge, including civil facilities, to meet both military and civil needs. A Military Transportation Directorate controlling military Transportation formations, he argued, was the normal way that this was achieved. It was crucial that a portion of executive transport personnel were disciplined troops who were trained in defence against ground troops, gas attacks and aerial assault. Military formations would also be able to be withdrawn more easily than 'unformed bodies of civilians'.[12] Biddulph outlined other cogent reasons for transport leaders to military officers. They had to be able to command military units. They had to conduct planning on how to meet the challenges of operational situations that would be dictated by the enemy, including withdrawal. Finally, in order to carry out planning based on military appreciations, they needed access to both 'secret' and 'most secret' papers.[13]

Unfortunately, the available sources do not explain a gap of nine days between Biddulph's call for militarization and its partial invocation on 25 February. Possibly, there were backdoor political manoeuvres that staved off final implementation of the militarization plan. It might also have been the case that this time was required for a wait-and-see period. The latter scenario might be more plausible in the unlikely case that Biddulph's proposals were a bluff to encourage civil heads to do more than they had. Whatever the reasons, the fact remains that the delay in militarization was delayed by another precious week and a half.

The impending militarization of the BR was no longer in doubt after what Biddulph termed the 'railway crisis' of 24 February.[14] As Dorman-Smith admitted, the breaking point for the civil administration of the railways in the south was reached by end of the third week of February. By this time, several hundred 'broken and fatigued' railway employees refused to remain in Rangoon and requested evacuation.[15] The situation was brought to a head by a telephone call by an unknown party to the railway administration. The caller claimed that the signal for the evacuation of everyone save the 'Last Ditchers' was going to be given at 1400 hours. In fact, that order was not actually going to be issued yet, but railway administration staff did not follow up the telephone call to confirm the message. Nevertheless, news of the call spread like the emissions of a suppurating sore. Later that evening, the lights went out in the loco sheds and signal boxes throughout the railway network as 'many scores of vital railway men' now abandoned their posts and simply went home, leaving 'only a tiny nucleus behind'.[16] So many left on the same night that Chance surrendered to supine and misguided fatalism. In the midst of a campaign in which the defenders would clearly have to depend on the BR, he declared, 'the railways are finished and not another train can run'.[17] The railways south of Toungoo had ceased to operate, except for several suburban and local trains run by officers and subordinate staff in the dock areas of Rangoon, as all staff had been evacuated or fled.[18]

Biddulph rejected Chance's assessment. His argument was that the railway was not free to simply stop working. It was essential that the railway evacuate troops, refugees and stores from Rangoon, on the one hand, and to keep the troops at the front in supply, on the other, especially since the railways were now the only means of communication between Rangoon and the fighting formations. It was essential, Biddulph asserted, that a minimum of one train be run on the Pegu line and on the Prome Branch each day. There were sufficient staff on hand to do this work. The BR still had twelve to fifteen railway officers, there was still some railway staff on duty and the men of the HQ, No. 3 RCMG and Engineers of the 17th Indian Division could work as traffic staff as well. Such a force would be sufficient to run a train on each branch if the engine crews themselves worked the points in the yards and in the docks' sidings. Chance refused to consider the proposals. Reportedly, Chance asserted for a second time that the trains could no longer operate. Chance's disposition was probably reinforced by the fact that on 24–25 February the Japanese renewed an intensive air assault on Rangoon with 150 bombers and fighters,[19] which, along with the military reversals in Tenasserim, encouraged the image that the Burma defence was now playing its end-game.

Chance's admission that the BR had ceased operations provoked direct military intervention into the BR administration. On 24 February, when it was clear that civilian railway administration had stopped cooperating and indeed that civilian railway operations had ceased to function altogether, the Burma Army informally took control of the railways from Prome and Toungoo with the hopes of getting a skeleton service into operation under the aegis of the military.[20] Biddulph was supported by the military command in Burma in undertaking this move. Goddard's report made on 29 March 1942, for example, probably represents the prevailing view in Burma Army HQ at the time. For a considerable period of time, Burma Army HQ had been frustrated by having to negotiate with many civil departments and agencies who were unaware of how quickly things moved in war, with the result being 'interminable committees, agenda and minutes which have wasted time and thwarted any attempt to get things done quickly'. Among these, some of the glaring examples were the civil railways and IWT agencies who could not be convinced that there needed to be at least some military transport units and unified control until it was too late and resulted in the railways' breaking down, which prevented the backloading of 'valuable and vital stores' in Rangoon which were now lost. Goddard explained that when Biddulph arrived in the colony at the end of January, there were no military transport units in place because 'the [civilian] head of the BR could not be persuaded that it was essential to have some military units'. Nevertheless, Goddard explained it would not have mattered anyway, for had Chance been so persuaded, it was unlikely that the railway staff would have joined any such unit. India was unfortunately unable to provide transport units when asked. The lesson was, he argued, that 'transportation must be organized for war'.[21]

For military Transportation to work, Biddulph required formal control of the railways. Chance's admission of failure and the Burma Army's subsequent arrogation of control over important sections of the railway together gave Biddulph his opportunity to remove the thorny Chief Railway Commissioner (CRC) from the colonial railway administration. Biddulph requested that Dorman-Smith formally remove the railways in Lower Burma from Chance's control and place them under military administration. Presented with a *fait accompli*, Dorman-Smith relented on 25 February and consented to the takeover. Dorman-Smith now issued an order handing over the railways in the south to military control. The Governor agreed to what at this point was only a partial militarization of the railways in Burma. The colony was divided into two halves, running through Prome and Toungoo, with the northern half remaining under civilian administration and the southern portion including Burma's most important city, Rangoon, and all branches stemming from the southern lines now under military control.[22] It was in this context of continued Burma Army defeats and severe challenges to the railways that Chance accepted failure, on the one hand, and Dorman-Smith agreed to a militarization of the railways in Lower Burma, on the other. As Biddulph had confessed to Manton in a letter dated 1 March 1942, he had not expected much of militarization; instead, it was an 'eleventh hour' effort made because of the collapse of the civil railways.[23]

Biddulph decided his next move based on the local operational context. However, it would be a move that would lead to years of complaints about his performance in the FBC. Biddulph ignored the precedence of seniority in railway administration appointments and made Brewitt, the Deputy Traffic Manager, Transportation, his Director of Railways, confirmed by an order from Dorman-Smith.[24] Interestingly, Biddulph had earlier put forward names of candidates for the ROC should it materialize in his discussions with Chance. The undated notes from this conversation, recorded in Biddulph's personal notebook he kept during the campaign, mention that Chance was firmly against Brewitt's appointment or of placing the men under Prendergast. Instead, Chance recommended Johnson or Proctor. He warned Biddulph that the chief railway officers would not serve willingly under either Brewitt or Prendergast.[25] Regardless of whose fault these interpersonal conflicts lay with, it was clear that Brewitt was an outsider (much like Biddulph himself) or at least was perceived as such by a high-level BR official. It is highly possible that Biddulph chose a capable man who himself was an outsider and hence in need of an ally outside of internal cliques among the railway staff. Certainly, it was made very clear to Biddulph that Brewitt was uncompromised by any loyalty to Chance; if Chance had to go, it was likely that Brewitt would bear no resentment to the changing of the guard.

Brewitt was given the rank of lieutenant colonel (acting) in the BAF. He was then put in charge of the Control Office in the Railway HQ as Director of the Railways over the militarized section of the railways not yet in enemy hands. The District Loco Superintendent of Toungoo, John Douglas Lewis, was given the rank of major. James Alexander Ferguson, District Locomotive Superintendent and Transportation Officer (Power) was made a captain, as was District Locomotive Superintendent Arthur Johnson, and A.F. Fitzeherbert for remaining at his post at Thaton. Burma-born Hector Soord was made a corporal. Brewitt and his chief officers went to Prome to manage the railways traffic as part of a new ROC. His other officers remained where they were at the time of militarization, Corporal Soord, for example, was now in charge of the withdrawal of equipment from the Locomotive Workshop at Insein, nine miles northwest of Rangoon, to the Carriage and Wagons Workshops at Myitnge, ten miles south of Mandalay. S. C. Bryant, the Deputy Traffic Manager (Commercial) from 1940, and the Deputy Chief Engineer C. C. R. Edwards both joined the militarized BR men at this time too. Reputedly one of the great minds of the BR, C. N. Blakeney, who before the outbreak of hostilities had been moved out of the Commercial side of the Traffic Department to become Secretary to the CRC in March 1938 and then Statistical Officer on 1 September 1940, was now included in Brewitt's military operating company as well.[26]

Milne was particularly irked that the Director of Transportation (Biddulph), the Director of Movements (Soden), and the General Officer Commanding (GOC), LofC (A. V. T. Wakely), focused only on the needs of the military at the expense of civilian traffic. As he observed more generally of all three military authorities with whom he dealt, none had Burma experience, nor did they understand the situation of outdoor staff or the reasons the latter were absconding. More broadly, Milne disagreed with military methods and priorities. He argued that a mistake had been made in allowing military men without local experience, who would also not accept the advice of those who had this experience and knew what could and could not be done, control all the forms of transportation in the colony. Dorman-Smith apparently shared Milne's views. Milne also championed the cause of the colonial railwaymen. He argued that the railwaymen were concerned about having to work the line while their families remained unprotected back in Mandalay. It should be a 'matter of urgency', he stressed, for the army to get these families out of Burma, the implication being that the BR men would not only work better as the war progressed northwards, but in fact it would help to prevent an all-out collapse of railway operations.[27]

Although reports at the time sometimes suggest that Brewitt commanded a formal military ROC, this is not strictly true. Instead, it was a nucleus Railways Directorate (Burma) such as that outlined by Biddulph earlier in February. For Biddulph, the more important part of Brewitt's mandate was that he was to recruit BR men into prospective military railway companies that would be formed later when the situation had stabilized. In the meantime, these men would be militarized while they remained in service on the line. Brewitt put together a group of men from the railway staff who had remained at their posts and then despatched them up the line on the Prome Branch to gain contact with

Map 4.1 Militarization of Burma Railways, late February 1942.

absconders and attempt to urge railway officers and staff to return. Nearly eighty men returned to their posts almost immediately. These men were now militarized and tasked with running the southern half of the railways, from Prome and Toungoo south, closer to the front lines. Above him were the military Director of Transportation (Biddulph) and the Director of Movements (Soden). Following their orders was a new railway operating battalion, agglutinated from existing civilian railway officers, now drafted and given military titles. Messages were sent to each station between Rangoon and Prome calling for volunteers to join the new and prospective railway military units. Burmah Oil Company officers also lent additional help by working a shunting engine from the railways to the Burmah Oil Company siding, taking loaded petrol wagons to BR sidings and returning empties for further loading.[28] The railways were now running again.

The railway staff who remained unmilitarized were sent to Mandalay. These included Chance and all of the insiders at the top of the railway administration such as the respective heads of the three railway departments, the Chief Engineer (Proctor), the Locomotive and Carriage Department Superintendent (Norman Johnson), the Deputy Railway Commissioner (William James Air) and the Traffic Manager (Milne). These men left Rangoon on 25 February in a motor convoy to Mandalay via Pegu, arriving there on 26 February to set up the new railway HQ. Mandalay Station was unable to accommodate the staff and office necessities of the entire railway administration that had now evacuated Rangoon. As a result, arriving offices and departments found new homes at several locations. One of the main relocation points was Ywataung Junction, some fourteen miles away on the north bank of the Irrawaddy across the Ava Bridge. The staff of the establishment, stores and commercial branches of the Traffic Manager's Office had already been partly evacuated there and now Milne re-joined them. It had been arranged that newer arrivals from other offices would be sent to Thazi Junction instead.[29] For the time being, the railway's extensive telephonic and telegraphic communications network made it possible to physically decentralize the railway administration in this way. This arrangement would not work so well in the last weeks of the campaign, however, when both telephonic and telegraphic communications had broken down. For unclear reasons, Milne and the Commercial Department did not remain with the other offices at Ywataung. While the latter seemed to be staying put, the Commercial Department was shunted around from Ywataung to Mandalay and then again to Maymyo.[30]

The most important railway asset that had to be withdrawn, however, was the railway workshop at Insein. Alongside the need for water and fuel to keep the engines running was the need to keep them in repair. The technical aspect of the railway was reflected in the layout of the BR' repair facilities. At distances of every 100 miles, the railway network included locomotive running sheds where minor upkeep and light repair work could be done.[31] However, major work could only be conducted at the two main railway workshops. The focus of the railway on the colonial export economy led to the concentration of these facilities near the two major ports of the colony. The first was at Insein on the outskirts of the seaport and colonial capital of Rangoon and the second was at Myitnge ten miles south of the river port of Mandalay. As the main hub of traffic and the point of entry for new railway equipment, Rangoon was the most important repair and maintenance facility. The workshops there were intended to handle the locomotive fleet. During a normal year in peacetime, the Insein workshops carried out heavy repairs on 100–120 locomotives, light repairs on twenty-five locomotives and intermediate lifting of around thirty locomotives.[32] The Carriage and Wagon Workshops at Myitnge were closer to the forested areas in the Shan highlands and elsewhere in the northeast. The Carriage and Wagon Workshops, as the name suggested, were intended mainly for the construction and repair of railway wagons and carriages. However logical this arrangement in peacetime, in the context of the Japanese invasion of the south, it exposed Burma's main locomotive workshops not only to easy aerial attack but also to capture should the Japanese land by sea or come overland through the southeast, of which the latter course did materialize. From the perspective of the Burma Army, which had expected an attack through Chiengmai, this positioning might have been wise had the Japanese used that route. However, when it was decided to abandon Rangoon, it meant that the BR's main workshops would fall into Japanese hands, while simultaneously depriving the railways of the means to keep itself in operation for very long into the campaign.

BR men were thus given the twin assignments of extricating what railway equipment could be removed and destroying what they could not. The relocation of any locomotive workshop equipment that could be moved to the existing workshops up north at Myitnge was assigned to Soord. This equipment would be crucial to the railway surviving as long as it did in the following months.[33] The importance of the railway workshop equipment, the scale of the relocation of equipment and the increasing proximity of Rangoon to the reach of Japanese forces meant that the railway workshop equipment had to be prepared for removal very quickly, removing their availability for several weeks. Although visible removal efforts were avoided as much as possible prior to the evacuation order being issued, so as not to cause panic, some measures had already been undertaken. Among these, Cardew and his team had already disassembled several key machines at the Insein workshops from 1 February so that they could be removed quickly when the order was given. From 19 February, many of the already dismantled machines were loaded onto railway wagons and on the following day, they were carried from the Insein Workshops 377 miles northward to Myitnge.[34] The equipment ultimately removed included seventy-five machines or fifty to sixty per cent of the Insein Workshop's machinery capacity.[35]

Biddulph was increasingly hard-pressed for more men to keep the railways functioning. The overburdening of the slim ranks of the remaining BR staff, even with the help of Prendergast's men in the No. 3 RCMG, began to threaten to hamper military supply.[36] Biddulph sent an urgent request to India for another officer of the IE and both engine drivers and signalmen. However, help from India would be delayed for another week. On 5 March 1942, a small detachment of an Indian ROC, including Lt. Raymond Taylor Carr, would be flown in and placed under the orders of Prendergast (as the Assistant Director of the Railways) to reinforce the railways staff in the militarized part of the BR. They would remain in Burma working alongside the BR men until they walked out of India with them two months later.[37] In the meantime, the situation had become dire. By the morning of 25 February, railway traffic throughout the network was at a complete standstill. There was no coaching or goods stock in Rangoon or in the railway yards nearby. Everything was at a 'complete standstill'. Among the first tasks was to get the railway running again in the south. The same day, BR officers who had been commissioned as part of a ROC and remaining subordinate staff were detailed for different duties and posted. Borrowing from other government departments made up for shortages of staff. Brewitt also made an appeal for a return of railway staff that led to seventy railway staff returning to work on 1 March alone.[38]

Burma Army officers also lent a hand at the stations in their proximity. One of them was James Lunt. His unit, the 2nd Burma Brigade, was moved to Nyaunglebin in mid-February. Here, Lunt remembered, he became something of an expert at shunting as he marshalled trains in the railway yard in the first week of March.[39] Another example is that of Prendergast. As mentioned earlier, Prendergast went to work mobilizing Indian refugees he found at the main railway station in Rangoon to do the necessary menial labour. With their help, he was able to muster at Rangoon Station sufficient empty wagons and a shunting engine to form a train. On 26 February, all of the wagons not carrying military supplies were cleared from Rangoon's Montgomery Railway Yard. The railway now got shunting work there started up again. The sidings at the Rangoon Docks were also cleared up of all obstructions and impediments to railway movement.[40]

Alexander's arrival and the railways

The railways had been busy for the several days preceding the arrival of Gen. Harold R. L. G. Alexander to take over command from Hutton. In an effort to delay the Japanese advance until his arrival, the railways in Lower Burma were devoted completely to the tasks of ferrying troops. Because the Japanese were crossing the Rangoon–Mandalay Trunk Line, no trains were running between Burma's two largest cities. A notice directed at would-be civilian passengers was pinned up at Mandalay Station at the beginning of the month announcing that all trains to Rangoon were temporarily suspended.[41] Instead, two trains a day were being run from Rangoon northwest to Prome.

On 4 March, Alexander arrived with expectations of saving the situation, but when Pegu was lost a few days later, he accepted that Rangoon now had to be abandoned.[42] It was now too late to evacuate by rail to Mandalay. If the Japanese were merely crossing over the railway and road communications between Rangoon and Mandalay, including the Rangoon–Mandalay Trunk Line, by 5 March, their encirclement of Pegu had absolutely cut both of them. Now the Japanese were crossing the railway lines and roads everywhere to the east and north of Rangoon and thrust west in an effort to trap Burma Army HQ in the colonial port. Advanced Army HQ, administrative units and any troops not assigned to protect the demolitions teams, however, moved out on motor transport on the Prome Road. Advanced Army HQ was to go to Maymyo and there re-join Rear Army HQ, Alexander having flown out to Maymyo just before Rangoon's fall to prepare for its arrival. The last empty southbound train able reach Rangoon Station was assigned the task of moving whatever part of the Burma Army HQ that had not left on motor vehicles. This train was under Soden's control, acting as the Officer Commanding the train. Only members of the Burma Army HQ with permits to leave on the train were allowed on the platforms at Rangoon Station.[43] Indian troops loading up military supplies in the port area stopped their work on the afternoon of 7 March and at 1300 hours they despatched the last goods train from the sidings at the Rangoon docks to Prome carrying 500 tonnes of supplies. An hour later, the 'D' (Denial) signal was hoisted, instructing the Last Ditchers to carry out the demolitions. Meanwhile, the Burma Army regrouped northwards in the Irrawaddy Valley.[44] The Japanese advance on Pegu meant that rail passage was now only possible to the northwest to Prome. The last trains heading out in this direction from Rangoon were the three trains that carried the Last Ditchers.

The colonial railwaymen seem to have been cognizant of the changing quality of their operational environment now. During the second week of March, when Brewitt's militarized railwaymen were dispersed between those parts of the Prome Branch and the Rangoon–Mandalay Trunk Line that remained in British hands, this group of militarized railwaymen found the changing demands placed on the railway and the flurry of administrative changes confusing. The War Diary for the 1st Burma Division HQ, the division holding its position on the trunk line, noted on 10 March that civil administration had broken down in Lower Burma as a result of effectively all government workers having left their posts and that the disorganization consequent to this was most evident among railway personnel.[45] Certainly, from March 1942, railway staff found the danger of their work had increased dramatically. Until the fall of Pegu, the Japanese had taken only the branch line which served Tenasserim and which could be abandoned without sacrificing essential services to the army and to China. However, after Pegu, the main railway system had been breached. The railways had to be kept running very nearly up to the front lines. As the front was highly mobile and usually moving north and west, railway staff often found themselves in continued danger of being killed or captured if British forces retreated too rapidly or the Japanese engaged in an encircling movement. Nevertheless, Brewitt would soon have them pulled back together. The militarized railwaymen would keep the trains going once he and the rest of his men had left Prome and further steps were undertaken to militarize the railwaymen in the Upper Burma as well.

Brewitt had already set up his railway HQ at Letpadan on the Prome Branch a few days before the abandonment of Rangoon. While remaining railwaymen worked on denial operations in Rangoon he ran a train service between Letpadan and Prome in support of withdrawal and then, on 8 March, ran three empty trains to Taikkyi on the line to Rangoon in order to fetch wounded and exhausted troops who would otherwise not make it to Prome.[46] At about this time, Brewitt penned a report, 'Military Railway Operation in the Defense of Rangoon, 25th February–8th March, 1942' to Biddulph on what his BR men aided by Prendergast's men had been able to remove from Rangoon. Working 'day and night', they had cleared out all of the military goods from Mahlwagon and then sent 200 empties to Mingaladon and Tadagale where they were also filled and despatched. In this week-and-a-half long period, the militarized BR men had saved 12,000 tonnes of RAF and other military supplies and supported numerous troop specials to the front and then to Prome. They had also run evacuation trains carrying 3,000 to 4,000 dismounted and exhausted troops for whom motor transport was unavailable from Pegu and Taikkyi out of the area and to safety, thus avoiding certain capture or death.[47]

Towards full militarization

After the losses of the last few weeks and in the face of the Japanese sweep across the delta, coherence to the fighting organization needed to be restored. The mass of soldiers and equipment collecting at Prome represented both an end and a new beginning for the defenders of the colony. Imperial forces received new leadership in the form of Acting Major General (soon to be Lieutenant General) Slim, formerly GOC of the 10th Indian Division in Iraq. Alexander decided to form the 1 Burma Division (at Toungoo at the time), the 17th Division (about thirty miles to the south of Prome) and the newly arrived 7th Armoured Brigade into a corps under Slim.[48] Slim assumed command of the newly organized I Burma Corps, or BurCorps for short, on 19 March. The BurCorps would include all British, Indian and Burmese forces fighting in Burma and would constitute at this point, if it had not done so before, an imperial force both in its command and in its rank and file. Just as the railway was under two officers brought in from outside Burma, so too was the army.[49] Alexander decided to divide the defence of Burma between imperial forces and the Chinese Expeditionary Force that arrived at this time under Gen. Joseph Stilwell, with BurCorps defending the Irrawaddy Valley and the Chinese Expeditionary Force the Sittang Valley and, with it, the railway to Mandalay.

Frugal colonial infrastructural policies determined the futility of any defensive strategy after the loss of Rangoon. Fewer ports meant lowered development investment and savings to the colonial state. This meant that when the Japanese captured Rangoon they had the only major seaport in Burma. Alexander's defence outline for Burma, after Rangoon, required the drawing of a new LofC for the remaining imperial forces in the colony. At the time that Hutton had ordered the redistribution of depots and reserves concentrated in Rangoon to other parts of Burma, Goddard had been asked to scout out a possible new port of entry as an alternative to Rangoon, if the latter were lost. Bassein in the western part of the Irrawaddy River was the best choice because of its location. However, colonial economic development had skewed the pattern of the rice export trade away from the development of Akyab (today Sittwe), Bassein and other ports in favour of developing central control over the flow of people and commodities by relying on Rangoon as the only major seaport. Bassein thus had no port facilities, such as wharves, cranes and the like, and it would take at least six months to build an adequate wharf there and a full year before the port could accommodate nine seagoing vessels at a time.[50] The next best option was to use Prome upriver, along the riverine artery connecting Mandalay to Yenangyaung and what remained of imperial and colonial forces further south. The Japanese meanwhile were able to occupy the lower delta, cutting the defenders off from maritime connections and lines of supply anyway.[51]

The pre-war transport regime in the colony played an important role in determining the Burma Army's decreasing reliance on the BR after the loss of Rangoon. As the Burma Army moved towards the Irrawaddy River and away from the Sittang, it would rely mainly upon motor and not rail transport. Geographical reality effectively bifurcated the colony into what were natural service zones, mutually distinct and exclusive, relative to transport means. In other words, the IFC could not be made to take over the role of the railways in the eastern half of the colony, as it required an overland system and the BR could not take over the role of the IFC in the western part of the colony, as it was dominated by a river system.[52] Withdrawal, reorganization and a new LofC meant that the railwaymen at Prome had to be redeployed as well. When Hutton left Prome for Upper Burma on 11 March, he brought Prendergast with him as his own preferred choice to take over command of the railway in Upper Burma, which remained to be militarized. The remainder of the No. 3 RCMG, except for a party of six, and a majority of the BR personnel were moved from Prome to Mandalay on 18 March. The detached party of six kept the Prome Branch going for as long as possible.[53] An air raid on Prome Station then simply stopped trains on this branch from running altogether.[54]

Although the bifurcated civil–military administration of the railway had another week or so of life in it, discussion had already begun in early March on the possible militarization of the railways in Upper Burma as well. In his private correspondence with Hutton after the loss of Burma, Dorman-Smith claimed that he had offered to hand the remainder of the railways over to army control when he arrived in Maymyo, which occurred on 2 March. The army, however, asked him to wait until Biddulph had left Prome, causing some delay to what was already, for everyone, a foregone conclusion. Finally, on 14 March Dorman-Smith, motivated by the changed circumstances resulting from the failing defence of the colony, took the legal steps necessary to end the split administration of the railway system by issuing a proclamation assuming all the powers exercisable by the BR Board, as well as declaring in this proclamation his intention that the railways should be under military control. Dorman-Smith did so under the authority he had assumed under the terms of the GOB Act,

1935. The BR Board was now suspended and what remained of the railway was militarized. The formal transfer of control would not take place until 29 March when Dorman-Smith delegated the powers he had assumed from the BR Board to the General Officer Commanding-in-Chief, Burma Army with the authority to redelegate.[55]

Dorman-Smith placed the management of the BR into the hands not of Prendergast, but of Brewitt who retained his title of Director of Railways, the title he had held in the Lower Burma since partial militarization. Brewitt and the other officers now moved up from Prome to Mandalay and then in April to Ywataung Junction.[56] The Governor also endorsed the raising from the BR staff of four BR military units, including the company organized in the south, to form the One Indian Railway Construction Company, IE; One Indian ROC, IE; One Indian Railway Workshop Company, IE; and One Indian Transportation Stores Company, IE.[57] The total number of BR staff who volunteered to join such units included thirty-four officers and 400 other ranks. Nevertheless, although the officers were commissioned and the men enlisted, no further progress was made with the actual organization of the units, although the men did now qualify for military rations and other benefits.[58]

On 18 March, some of the other senior railway administrators who had just recently relocated to Upper Burma, including the highest-ranking civilian railway officials, were effectively dismissed. CRC Chance was informed that he was no longer needed and, after finishing some reports on railwaymen for the BR personnel files, he took two years leave preparatory to retirement and the next flight out to India.[59] Rose, the Superintendent of Stores, was told the same thing; he then flew out. On the same day, Biddulph held a meeting with a small group of other senior railway administrators consisting of the Traffic Manager, Milne, and two other heads of departments, Proctor and Johnson. They were asked either to leave as well and seek employment in India or to remain and serve in a civilian capacity under Director of Transportation Biddulph. They agreed to the latter role, but not immediately. Although Milne would ultimately remain with his fellow railwaymen until the very end of the campaign, he at first was resigned to following Chance to India. His subordinate in the Commercial Department, Edward Law-yone, who reported for duty when Milne was still at Ywataung, stayed loyal to him to the latter's apparent end with the BR. Milne, who had risen so quickly in the late pre-war years under Chance, now languished in a Biddulph- and Brewitt-dominated railway. Over the course of its repeated relocations since the evacuation from Rangoon, the Commercial Department's staff gradually disappeared and these moves now merely amounted to the shifting around of Milne personally. Law-yone's entries in his diary make clear that he was on a friendly basis with his boss. Indeed, Law-yone provides a sympathetic picture of a lonely man now being moved about, the entire Commercial Department of the BR of which he was head was now reduced to two, Milne and Law-yone, and all of the other railwaymen had joined up with Brewitt. In fact, Law-yone held a dim view of Brewitt's militarized railwaymen generally, complaining that they were a cheerful, well-fed bunch who left Law-yone to struggle to do his job in addition to theirs at Myohaung Junction from his bombed-out carriage in early April.[60]

Milne apparently changed his mind about leaving Burma and decided to remain. The main task facing him and other staff in the north was to clean up the mess of a railway administration folding in upon itself. This included settling the affairs of the old BR, evacuating surplus staff and any help they could extend in the recruitment of lower ranking staff for the new railway operating and maintenance formations that would be raised and would come under Brewitt's command.[61] Milne's tergiversation on the matter of leaving the colony was no small concession given their history, but in this crisis bonds of loyalty among the BR men overrode personal rivalries.

Railwaymen in Upper Burma who had not yet joined Brewitt's group were now asked to volunteer. Thirty-four years old and a ten-year veteran of the BR, Assistant Locomotive Superintendent G. H. Binnie now joined. Deputy Locomotive Superintendent John Horace 'Jack' Cardew and other railwaymen at the Myitnge Workshops volunteered to join by 27 March. This required filling out the same kinds of forms as they had, in the case of First World War veterans, in 1914, so they could formally rejoin the army. The age limit was placed at fifty-one years, but apparently, this restriction was not observed given the circumstances. Supplementing the militarized BR men was again the No. 3 RCMG. The group was moved to Ywataung at the same time as Brewitt had been. From 3 April until 4 May, they would assist in working for the BR as air raid sentries, as guards and in the supply of rations to railway workers. As the centre of gravity of railway operations on what remained of the railway line pushed north, the group was bifurcated into one party that was shifted north to Kanbalu and another party under Busher who stayed at Ywataung. Busher's party remained in place until the last train had left and they then rejoined their comrades who had earlier been evacuated to Kanbalu.[62]

Feeding railway staff was an increasing problem because of the worsening military situation and the fact that Japanese bombing simultaneously increased over the entirety of the railway network. Spread out thinly at scattered stations spanning the course of the entire railway network, the railwaymen were having difficulty purchasing their own food locally because they were not being paid regularly. By 10 March, the railway staff had received only two weeks' pay in two months and as they were required at the stations for continual work, they were not to be running about trying to scrounge up food on their own. An even greater problem existed for engine drivers, who had to drive many hours without a break long enough to obtain or cook food. This situation worsened as the FBC wore on. By April, Law-yone, then at Myohaung, gave the go ahead to his assistant, Sergeant Marshall, to start breaking into stuck freight cars to forage for food, winning for the hungry railwaymen dog biscuits and dry sherry.[63]

There had been an expectation that militarization would have brought not only death benefits and authority to railway staff, but also the order to daily railway operations that civilians associated with the military way of doing things. This did not materialize. What was not realized perhaps was that while militarization may have aided railwaymen in operating the railway, it did not bring additional manual labour to find, cook and serve the food or replace the now-absent army of Indian sweepers, cleaners, washers and others who gave the coaches, the railway stations and other facilities the image of clean and orderly functioning. Stations empty of staff thus saw a breakdown of daily order and sanitation and things worsened as trains were increasingly held up at stations; passengers and staff eventually overcame the stations' normal resources.[64] In many stations, the latrines filled up and then defecation and urination took place in the spaces of the railway yard between the tracks and eventually throughout the precincts generally. The 'squads of sweepers' who would have been necessary to clear the filth were not usually available. As Prendergast complained, 'The stench became dreadful and it was very unpleasant for anyone walking about a yard', but admitted that it would have been difficult to stack dumps of lime in advance to deal with this problem.[65] On 1 April 1942, Maj. Colin Metcalfe Dallas Enriquez, the Assistant Commandant of the Mandalay Battalion of the Burma Military Police, complained that the army taking over the trains did not do much to get things in good order, Mandalay Station being filthy and the un-swept first-class carriage he was in having litter scattered everywhere.[66]

To be fair, however, labour shortages and the breakdown of civil order as the front rolled back meant that the disorder and dilapidation of the railway stations and carriages was true of the colony generally. Such highly visible and fetid problems hurt confidence and morale. It also contributed to an overall impression throughout the colony, or rather those parts served by the railway, that both the defence of Burma and the colonial order in it were failing. Indeed, Chiang Mei-ling recalled that Generalissimo and President Chiang Kai-shek informed Alexander after the couple's visit to the Burma Front in mid-April 1942 that if conditions like these had taken over in China 'heads would have been chopped off'.[67] The stench of the railway stations thus carried with it the smell of an impending military defeat, the collapse of a government and a quietus on a way of life.

Chapter conclusion

Biddulph and his men kept the railways working and war materials, soldiers and civilian evacuees moving in Burma under some of the most trying conditions perhaps any railway had had to face up to that time, with the exception perhaps of the Eastern Front in Europe at the time. As this feat was undertaken after the civilian railway administration had stopped functioning, this was indeed a major accomplishment, as Biddulph suggested in his letter to Manton on 1 March.[68] But as well as the railways did under the conditions, there were factors that continued to weigh against its performance that could have been avoided had imperial and colonial relations been more intimate and the circulation of metropolitan military railway operations knowledge been treated as the imperial knowledge it needed to be once war had broken out in Asia. The failure to do so slowed the pace, diluted the completeness of the militarization of the railways in Burma and hindered the full exploitation of its advantages.

Everything from the oft-cited 'fog of war' to exaggerated Chinese communiqués of victories against the Japanese in Burma also delayed a realization of the full scope of the Allied defeat in Burma in 1942.[69] But once Burma was lost and both participants and observers began to assemble, weigh and debate the reasons that explained the loss of Burma in 1942, the lack of military experience among many of the BR men was one of the factors that would be cited. Militarization had been viewed as advantageous for providing not only authority in handling railway services, but also for making it easier for railwaymen to risk battlefield railway operations because there would now be a guarantee that if they were killed in the line of duty their families would receive financial support. What this thinking missed, however, was that operating a railway in a combat zone was not simply a matter of personal risk, however much that was on any railwayman's mind, but also involved severe operational challenges, for which training would have helped considerably. Perhaps more importantly, training of some kind could have also reduced the friction between imperial military officers and colonial railwaymen. This friction was not due to essential conflict between a military and a civilian way of doing things, but instead a clash of operational cultures that would have resulted from the bringing together of men from any two different railway cultures, military or civilian, into the management and operation of any given railway system. Whatever the specific challenges, for the BR men to have received no training in operating the railways in war conditions by February 1942 was appalling by any standard. The reasons for the delay in doing so were, again, the obstacles placed on successful preparations by the limited vision of colonial authorities and the tensions between colonial and imperial military authorities.

5

Kings of the road: Decentralization and local initiative on the railway

The railway administration in colonial Burma was a small departmental system stretched over a very large territory, populated by a labour force that, while not miniscule, was small enough to allow a European incumbent considerable autonomy from the main administration. As with the overstretched civil administration, local railway authorities were able and expected to exert considerable autonomy in decision making and implementing central policies. This economy of administration helped the colonial administration keep revenues flowing to Rangoon and then to Calcutta and London at the least expense and it helped to keep the BR as much out of debt as it could. Sometimes this autonomy afforded and indeed required significant leeway to engage in activities that benefitted the colony and the larger imperial project, as Atsuko Naono has shown in the case of the experiments regarding smallpox inoculation at Meiktila.[1] It did run the risk, however, of corruption in outlying areas as Jonathan Saha has recently shown in the case of the Irrawaddy Division.[2]

One legacy of local operating autonomy was that while much of the railway staff were prone to flight, a handful of local railway managers, in particular district superintendents, not only remained at their posts but were able to run local sections of the railway without regular orders from the railway headquarters (HQ). This was crucially important to Biddulph's successful resurrection of railway operations in late February 1942 at a time when telephonic and telegraphic breakdown and physical isolation from Rangoon during the period meant that large sections of the railway were no longer in regular communication with any central authority. This chapter examines these developments in the context of the section of the line running south from Mandalay to Toungoo, what used to be the main Rangoon–Mandalay Trunk, clipped off now by the Japanese invasion.

The key figures in railway operations on the lines south of Mandalay that remained in British hands included Maj. Duncan Coward, the newly militarized Major Lewis, and Lieutenant Carr. Nevertheless, their roles have remained murky until now due to the poor documentation available regarding railway operations in Upper Burma between February and April 1942. These men were responsible for keeping the railways working in this section during the period of the first 'dance of armies' that eventually introduced Chinese armies and even American commanders to the BR in April 1942.

Coward, Lewis and the 1st Burma Division

Early Prussian observations of the speed and flexibility gained by Napoleonic armies by communicating quickly the general's intentions and rationale to commanders and allowing them their own initiative in realizing these led to the evolution of *Auftragstaktik* in the German Army. As military historians point out, the shift from set piece battles to conflicts of manoeuvre, speed and geographic dispersion made it difficult for commanders to develop up-to-date appreciations of the tactical situation. A system that relied only on orders to provide direction was not only inefficient, but also ineffective. What became 'devolved command', 'mission command', or 'mission-oriented command' was a command philosophy essential for Manoeuvre Warfare that allowed rapid choices to be made in rapidly changing circumstances and the taking of decisive action without the delay caused by waiting for orders from the top. The British Army realized the tactical benefits of decentralized command by the First World War. Although in the Second World War, the British Army would have two different doctrinal strands in its published tactical doctrine, as Christopher Pugsley has explained, both would espouse decentralized command. At the same time, the US Army during the Second World War was doctrinally committed to the 'command-driven' (*Befehlstaktik*) approach.[3] Nevertheless, as Patrick Rose has recently argued, American commanders, like their British comrades, were able to use the essentials of mission command like the Germans, to defeat the latter, but admittedly had followed a different pathway to achieving these successes in Italy by the end of the war.[4]

Just as devolved command was important in Manoeuvre Warfare, decentralized authority would be crucial for running the railways under especially disruptive war conditions. The benefits would be realized in the running of the BR during the FBC. Certainly, the nature of the time and circumstances means that how much of decentralization was forced upon Biddulph or something he would have pursued anyway cannot be fully discerned. Nevertheless, in his writings afterwards he identifies decentralized authority as a key part of successful Transportation Technique. It is also the case that Biddulph had planned to delegate authority to small ROCs working sections of the line autonomously after he deposed Chance and made Brewitt director. He hoped to have local, militarized groups of railwaymen to operate on their own regarding local conditions in pursuit of central directives. Just as the tolerance or encouragement of local initiative through command mission philosophy made more efficient and effective local military circumstances, the decentralization of authority proved crucial for continued railway survival in the FBC. As in the pre-war period, the problems of running the railway during the FBC were not uniform across the network nor were they static in any one geographical area of railway operations. Railway operators responded to different demands that changed in quality and intensity as the campaign progressed, with performance that was determined by the changing reserves of manpower, motive and fuel resources at its disposal, and these concentrations were heavily localized and ephemeral. They required local experience and local assessments to manage.

The most dynamic and challenging demands put on the railways during this period were presented by the rapid turnover of major military users of the railway. The details of the 'dance of armies' dealt with in this chapter are as follows. The 1st Burma Division moved south from December 1941 to the end of February 1942 to meet the Japanese on the Rangoon–Mandalay Trunk Line, its place garrisoning the Shan States lines of communication (LofC) was taken up by divisions of the Chinese 5th and 6th Armies. These troops then started to take up positions to the rear of 1st Burma Division. The first groups of Chinese soldiers were moved from Lashio to Pyu and then to Toungoo. The first contingents whose move was completed by 15 March consisted of 11,600 Chinese troops. These troops then took control over the Pyu–Toungoo area from the 1st Burma Division as the latter was withdrawn and moved to the west, to aid in the defence of Yenangyaung and Central Burma.[5]

Coward was the first of the three to deal with the 1st Burma Division, under Maj. Gen. James Bruce Scott (1892–1974). Coward had been made the Railway Transport Officer (RTO) at Thazi when the war with Japan broke out, which put him in control of the vital junction that connected Shwenyaung with the main trunk line that ran south to north, from Rangoon to Mandalay. When a possible thrust by the Japanese across the Sittang became a possibility, the 1st Burma

Division was moved south off the Shan States LofC, most of the division coming down by rail from Shwenyaung, although the Punjabs and Rajputs marched from Taunggyi to Thazi on foot. Other elements of the 1st Burma Division came down through Thazi as well, including the 7th Burma Rifles from Mandalay, Brigadier Jones's Independent Infantry Brigade, also from Mandalay, and some units of the Frontier Force that came down the line from Lashio, Meiktila and Pyawbwe. Coward's first major task was to ensure smooth rail passage of all of these units down the line through Thazi Junction in order to block any possible Japanese advance north into the heart of the Burma lowlands, especially the rice basket of Burma at Kyaukse, Burma's hot season capital, the hill station of Maymyo and Mandalay, Burma's second city.[6]

By contrast to the early flow of traffic down the line, developments in February saw the flow of traffic redirect completely. As mentioned earlier, Thazi was identified as the overflow centre for those railway offices that could not be accommodated at other railway locations in the north. During the two-week period between 22 February and 5 March, when the Japanese cut the line at Kyatkon, with the exception of a mere two downward trains – the first bringing empty goods stock and the second an empty ambulance train – nearly all rail traffic moved up the line and consisted solely of evacuation specials, troop trains, ambulances and military stores.[7]

Communications breakdowns for long stretches of time meant that local railway managers increasingly found themselves managing local stretches of the line on their own. According to Coward, all communications between the railways in Upper Burma and Rangoon had been cut even before the Japanese had crossed the Sittang. The major break in communications between Rangoon and Mandalay came on 22 February when the Japanese bombed Toungoo for the first time and destroyed completely all telephonic communications. Although repairs would be attempted, declining railway movements, the withdrawal of staff from Rangoon, the flight of railway personnel north and the Japanese advance northwards in close proximity to the line meant that communications between Rangoon became intermittent at best and then stopped altogether. The breakdown in communications had initially led the CRC, Chance, who was on his way to Mandalay, to ask Coward to assert his own control over the railways in Upper Burma. Coward's management would continue until the third week of March 1942 when both the civil officers of the railway administration and Brewitt's militarized company arrived at Mandalay.[8] The attention given to railway operations in Rangoon and the south obscures railway operations in Central and Upper Burma during this month-long period between the end of

February and the end of the third week of March. The latter period is not well served by the loss of railway records during the retreat out or the fact that most attention was directed at Brewitt's ROC. Nor did the main figures in the colonial railway administration devote much space in their memoirs to their experiences in their first few weeks of exile to the north, in part perhaps because they were too busy sorting out a new administrative scheme for railway operations that would not be centred on Rangoon. Aside from the occasional references made by military units such as Scott's 1st Burma Division when it relied upon the railway for its failed drive south, we had until the discovery of the Biddulph papers no detailed account by the railwaymen themselves of railway operations in the north during this period.

There are clear historiographical limits to relating the events of this period and place. The availability of Coward's and other accounts in the Biddulph papers goes far in giving some of the major themes, but a detailed narrative remains at present beyond the historian's grasp. The main problem here was that daily railway operations normally provided no sense of chronological change. Chronology remains the bedrock of history but it requires change to give meaningful markers. Routinized daily activities do not lend themselves to producing chronologies although they are helpful in characterizing periods and activities. Worse, for a colonial railway in a wartime situation, the overworked group of railwaymen were at their posts around the clock, which, Coward observed, made it difficult to keep track of the passage of time.[9]

The Japanese advance north up the railway line towards Toungoo made the railway critical to the defence of Burma and the delay of the Japanese arrival in Mandalay which would have trapped substantial numbers of soldiers and evacuees south of the Irrawaddy. Toungoo had the largest railway facilities in this part of the network and was under the authority of the District Locomotive Superintendent at Toungoo, Lewis. Lewis faced a number of challenges to keeping operations going from the last week of February. One of the worst was a shortage of labour. A Japanese air attack on 22 February had prompted the flight of most of the railway staff. The small number who remained at their posts included the European railway officers, Lewis and H. C. Palmer, and a few Indian and Anglo-Indian staff. Another was the physical damage to the railway tracks and facilities. After the 22 February air raid, Lewis personally dealt with three unexploded bombs in the railway yard, but the air raid on Swa twenty-five miles up the line the following day completely destroyed the tracks in the station. Lewis and his men had to cut out the station and eventually ran one line through. Yet another problem was looting by villagers along the line that increased in the escalating chaos invited by the invasion. Lewis wrote of the fears by railway staff of looting north of Toungoo after Japanese bombing of Swa on 23 February and their request for military guards. Matters worsened on 27 February when Toungoo was bombed again in a raid that burned down seventy-five per cent of the town and destroyed the bazaar. The remaining staff could no longer obtain food, were no longer being paid, were frightened and thus fled.[10]

The strategic importance of the railway to military operations soon brought some military assistance. On 1 March, Scott, the commander of 1st Burma Division, met local BR managers. He instructed Lewis and Palmer to get the Toungoo railway station open again, with Lewis in complete charge of railway operations at Toungoo. The railway in this section was now directed entirely to serve the needs of the 1st Burma Division. Scott communicated these needs to Lewis in daily conferences.[11] To support Lewis, Scott agreed to provide military rations for essential railway personnel. The Field Cashier now supplied money to Lewis to provide one month's pay to staff who stayed and to those who returned and promised that subsequent pay would also be handled in the same way as necessary. Lewis noted that conditions improved quickly, numerous staff returned, and a survey was done of their ranks, determining who was necessary and who was not, sending the latter up north to Yamethin and Myitnge where their disposal would be determined.[12]

Lewis attempted to put the railway north into better order with the limited resources he had at his disposal and those he could make so. He telephoned the Traffic Manager who gave instructions to commandeer wagons containing foodstuffs in the railway yards. Whatever was necessary to feed staff was retained, while the bulk was handed over to Supply & Transport. Lewis also telephoned the Chief Railway Engineer at Mandalay on 2 March requesting that railway officers be sent to Pyinmana, Thazi and Yamethin to keep up the morale of railway staff there and thus ensure that essential services were maintained. He also requested of Scott that military officers be sent to all stations where they would help determine the destinations of goods wagons and thus clear up congestion in the yards. On 3 March, he approached the colonial civil authorities about dealing with the problem of refugees in the stations and on the lines. Lewis argued that crowded trains, crowded platforms and people walking on the railway line itself made it impossible to conduct shunting without killing people. One solution, he proposed, was the erection of camps where refugees who could not be withdrawn yet could form a pool of labour that could be drawn from in exchange for food and free rail travel to Mandalay. This proposition was taken up and was responsible for getting numerous evacuation specials out of the station.[13]

While the railways had thus far supported the reinforcement of positions or withdrawals in the face of the enemy advance, in early March, the Burma Army used the railway to advance back into territory that had already been abandoned or taken by the enemy. At the end of the first week of March, when Rangoon fell, the railway activity furthest south on what remained of the Trunk Line was that of Lunt and others who were hard at work marshalling downward trains to send north. On 8 March, however, Commander Scott required the use of the railway to move his forces south into the section between Nyaunglebin and Pyuntaza. Scott's plan was to grab all the sugar and rice he could from the area and deny this to the Japanese. However, most railway staff at Nyaunglebin had fled and the railway south of the station was packed with deserted trains, not to mention the enemy. Among the few remaining railway staff Scott found at their posts were Palmer and Lewis.[14] How much Lewis had to say about staying at his post is difficult to ascertain. The regular army accounts refer very much to a man remaining stubbornly in the field, while the confidential report in Lewis's personnel file mentions that 'his services were more or less commandeered by Major General Bruce Scott'.[15]

When Biddulph introduced Transportation Technique into Burma and then militarized the railway, he had understood the limitations of civilian staff, even when militarized. Nevertheless, military commanders in the colony, many of whom were not versed in Transportation Technique, additionally were not aware of the limits of what or how much they could ask from civilian staff. The recently unearthed account of BR man Duncan Coward, included in the Biddulph papers, sheds some insight into why railwaymen were averse to supporting the military in the Forward Area. Coward, who will be discussed in more detail below, took charge of the Toungoo section of the railway in March. In the months after the Campaign, Coward wrote to Brewitt his account of what he had been told by one of the senior railwaymen who had remained in place in the area: that Scott had promised the railwaymen in this section that no operating company would have to serve closer than eighty miles from the front. Although Coward thought this was impractical and very openly did not honour this promise himself when he took charge, he confirmed that such a promise had been made.[16] When the military broke this promise in early March, railwaymen refused to take trains south of Toungoo towards the front. They also complained about running engines in the Front Area for the remainder of the campaign. Royal Engineers and others would bemoan the additional burdens placed upon them in early to mid-March by the lack of local civil support. This refusal was understood by elements of the military as evidence of a lack of willingness among civil operating staff to support the defence of the colony. As a result, military histories and accounts of the actions at Toungoo wrongly highlight the initial flight of railway personnel in early March and the refusal of railway personnel to drive trains south of Toungoo up to the front afterwards as examples of cowardice rather than as resentment of military assurances being broken.[17]

It also needs to be stressed that the military had not yet lent these civil railway operators any resources. While railway operators in southern Burma had been militarized and benefits were available to the families of those killed in action, the railwaymen in the north would not be militarized until mid-March. They thus had to worry more about the welfare of their families than did, for example, the military engineers who now had to drive the engines instead. Another complaint was that as they were not militarized they did not have access to army rations. This was a major reason why Pyinmana Junction seemed unstaffable – the town had been so completely evacuated that no rations remained. This was only resolved by sending a RTO on a weekly run to Meiktila with a lorry to bring back food. After the 1st Burma Division left the area a few weeks later, no rations were sent down the line at all beyond Pyawbwe, which would be a problem for Chinese troops there, discussed further below, as well. The local railwaymen complained that they had been promised that once the 1st Burma Division's move was completed, they would be relieved but were not. Finally, Coward himself noted that as the Japanese bombed the railways between 0930 and 1700 hours, the railway station 'always the main objective', he found it difficult to get the railwaymen into the yard to do work during the daytime.[18]

For the first few weeks of March, Scott thus had to rely on his own resources to get the railways back into shape in the Forward Area. The BR men other than Lewis refused, again, to go further south than Toungoo, not even into Pyinmana Junction. Scott thus had to turn to the two engineer companies attached to the 1st Burma Division, the Malerkotla Field Company, Sappers & Miners and the 50th Field Park Company, Sappers & Miners to get the lines open again south of Toungoo in support of Scott's advance. Lewis and the 50th Field Park Company cleared the line, built bamboo bridges sufficient to support the slow passage of a train, got the power-houses running again, erected ramps to load trains and marshalled the railway yards; where trains were jammed together extensions were built to circumvent those parts of the line and the lines were soon completely open to movement. Scott also had to find men who would fire the engines and drive them up to the front lines. Lewis lent his help to the 50th Field Park Company engineers to drive the troop trains.[19]

Scott assigned to the Malerkotla Field Company the task of shuttling back stores from Toungoo using three railway trains. The Malerkotla Engineers thus formed three crews of four men each. They also had men among them with experience of sorts in driving steam engines. One of the four-man crews was placed under Nk Sardar Ali Shah. Shah was a qualified steamroller driver and was now put to work operating the tactical trains that went with the 2nd Burma Brigade. Maj. E. R. B. 'Roy' Hudson, who had spent part of his youth riding in the cab of a local steam engine from Brookwood to the Bisley Camp rifle ranges in Britain and having thus acquired in passing a working knowledge of the operations of an engine, also now volunteered to drive an engine with himself on the footplate. Hudson and the two other Malerkotla Field Company crews took other engines down from Toungoo as far south as Nyaunglebin. Among the challenges on this stretch of line were the groups of Indian refugees who threatened to swarm the train if it stopped. Hudson's engines brought rakes of rice, sugar, stores and refugees up to Toungoo where the BR men took over and brought them north. Hudson was also tasked with hauling away empties, as many as forty a day, to the north partly perhaps for service, but also to clear up the line further. Scott's attack on the Japanese on 11 March failed to halt the Japanese advance, but it did succeed in getting much of the rice and sugar out of stations on the railway south of Toungoo. Nevertheless, it was not without human cost and two major head-on collisions left behind a grim reminder along the tracks of the dangers of running a railway close to the front lines.[20]

Japanese bombing of the line and the stations, in particular Toungoo, which was subjected to two bombing raids per day in mid-March, had a direct impact on the willingness of railway staff to remain at their posts.[21] Lewis's assessment of the impact of the bombing on the condition of the railway line and on staff morale led him to conclude that continuing to use Toungoo as a base of railway operations in the area was no longer feasible. Remaining in the smouldering ruins of the old facility risked the 'wholesale desertion' of railway staff. On 19 March, Lewis sent the railway staff north to Yamethin along with all of the engines and stock they were able to take with them, the convoy forming three trains. Lewis wrote in his report the same day that there were 'no railway staff resident in Toungoo'.[22] Interestingly, Lewis explains of his plans that he intended to keep doing at Yamethin what he had been doing at Toungoo. This was, he writes, 'taking charge', which conveyed his acting on his own initiative where clearly leadership had to be asserted but was not.

Among the arrangements Lewis made was the agreement by the military that the civil railway staff would only take trains down from Yamethin to Toungoo and only when necessary. Even then, the railway staff under him, described at the time as 'somewhat out of hand', resisted taking the trains down to Toungoo. Lewis himself would only remain at Toungoo until the 1st Burma Division had withdrawn. When that move was completed, he would move to Yamethin as well. At Yamethin, the Loco Foreman Green had recently lost some staff who had absconded, but he was able to more than make up for them by the wagon loads of railway staff who now came north, both the new arrivals from Toungoo and other refugee railway staff and their families evacuated from stations all along the line to the south. One military engineer saw fifty–sixty wagonloads of refugee railway staff and their families in the Yamethin Railway Yard on 20 March. There were more staff than they could use at the time at Yamethin. Major Lee, a Deputy Traffic Superintendent of the BR and one of Brewitt's militarized railwaymen, was thus up north at the time making preparations to send many of these refugee railway workers further up the line to different stations that were depleted of staff. Lee then had the empties sent back south again to Yamethin. There were some problems with refugee staff because these had not been paid, unlike those at stations such as Yamethin that had not yet been bombed out.[23]

Lewis had insisted that any railway work south of Toungoo would have to be conducted by military personnel only, under military orders. This was possible because of Prendergast's timely despatch to Toungoo of Lt. Raymond Taylor Carr of the East Indian Railway (EIR) Engineers. Carr had arrived in Burma with the 139 ROC, (Royal) IE on 5 March, just before Rangoon fell, and had joined Prendergast in the withdrawal out of Prome up north to Mandalay. After arriving in Mandalay, Prendergast had sent Carr down the line to assess the status of what remained of the main trunk line and the status of railway support in the Front Area. Carr and his men, including six he had acquired at Prome, followed him to Yamethin where they were put up in local rest rooms, those for enginemen, for British and for Indians, taking the precaution to dig a trench for themselves outside of the railway village, but within reach in case of an air attack. For the next few days, Carr's men spent their time 'learning the road' and gaining a good deal of experience at actual shunting in the railway yards at Yamethin. Lewis wanted Carr to use his own IE men to do work in the Railway Yard at Toungoo and to take the trains even further south. Lewis hoped these men would come down from Yamethin as far as possible below Toungoo, sleeping on the train, so that as soon as they arrived they could begin working.[24]

Carr maintained communications with Prendergast by writing occasional reports to the latter. He based himself in Yamethin at the end of the second week of March and went on tours of the line. He used these lengthy journeys as opportunities to write out his reports to Prendergast on what he had done and seen and on the overall state of the railway south of Mandalay.[25] The main problems facing train operations were due to staff reluctance to serve due to a lack of ration provision and bombing.[26] Staff were said to hide in the jungle during the day and only return again at night. W. H. Newton commented that on the way to Shwenyaung on the night of 20 March, a number of stations had no staff at all and at Thazi Station 'filth on the platform was indescribable'.[27]

The situation on the Toungoo Branch changed dramatically from mid-March largely because of the decision to divide the defence of Burma into an imperial and a Chinese zone. This meant the withdrawal of Scott's 1st Burma Division north out of Nyaunglebin over the course of 19–24 March through Pyinmana Junction and on to Taungdwingyi,[28] and its replacement by Chinese armies that now began to arrive on the Toungoo Front. With the change in armies also came the coincidental change in local management of the railways. Coward replaced Lewis as manager of railway operations in the section. The railway operations that Lewis had put together when he took charge of the section began to fall apart and would have to be reinvented by Coward. This was a consequence of the highly personalized nature of the local railway management. Local railway managers needed to keep the railway functioning in their sections. Lewis had created his local operating 'team', gathered the necessary equipment and got the railways in the section running again, in broad keeping with the overall goals and vision of the central railway authorities. These were in keeping with the general goals of the Transportation Directorate, which, above all, needed the railways running in support of the army. How Lewis realized these goals was left, not only by necessity but also by design, as Biddulph refused to assert local, direct management of railway operations, entirely to his own initiative, without orders from railway HQ or communications with the Transportation Directorate.

The railways' linguistic aspect and the Transportation of the Chinese Army

In terms of the railways' performance during the campaign, one of most damaging aspects of colonial-era railway development in Burma was its alienation from Burma's gargantuan neighbour, China. The railways had been inspired in the mid-nineteenth century by the possibilities of an overland route to Western China through Burma and an attempt had been made to build a rail link to China.

Although halted in 1902, it would be resurrected with efforts in 1941 to build yet another overland rail connection between Burma and China to complement the Burma Road. While no such link was made before the war, the railway branch legacy of this effort was the Lashio Line, and Lashio had emerged by the end of the 1930s as a major entry point for Chinese trade and people into Burma. This was much to the chagrin of colonial authorities. They sought to keep the Second Sino-Japanese War (1937–45) as far away from Burma's backdoor as possible. Nevertheless, years spent as part of the IR (until 1937) meant that they could sport alongside English and Burmese fluency one or another Indian languages, from Hindustani to Urdu, but according to surviving personnel files, not a single railway officer had familiarity with the Chinese language. During the campaign, several Chinese armies were called upon to enter Burma and fight, taking over the section of the front serviced by the railways while British forces defended the western side of the Irrawaddy Valley. This put the BR in the awkward position of being a railway operated by Chinese Army Officers who generally did not speak English. It was only the Chinese Liaison Officer (CLO), many of the officers of which strangely, despite their title, did not speak Chinese either, and a handful of American officers attached to the American Chinese Liaison Mission (CLM), who could help mediate many difficult situations.

Biddulph's instructions to Transportation officer Capt. N. W. G. Brown for the latter's liaison conference with Brewitt on 27 March laid out what the railways' relationship with the next Chinese Army contingent should be and what elements of the railway would be responsible for the move. In general, Biddulph stressed, it was important for Brewitt to understand the vital urgency of the importance of moving the Chinese forces as far south by rail as possible. The IE could handle much of this work. Carr's men, Biddulph expected, would be able to supply the necessary engine crews and, if necessary, the station crews as well. Brewitt's men, it was indicated, would need to ensure that at least one railway station would be open forward of Pyinmana and two trains had to be permitted to be forward of that station at any one time, about every two and a half hours from the Forward Area.[29] The next Chinese Army contingent would take the rest of March and into April to move down into position and consisted of about 26,000 Chinese troops who were to be moved from Lashio to Kyaukse, Thazi, Pyawbwe and Pyinmana.[30]

On the surface, the arrival of Chinese divisions on the Toungoo Front in March led to the effective division of the railways into two simultaneous railway regimes. Chiang Kai-shek had insisted that the BR be put under at least joint control under the British and the Chinese. In normal conditions, this kind of arrangement would have required formal deliberations on the division of powers, the identification of responsible authorities, debriefings and training that would have required months to do properly. The rapidly deteriorating situation in Burma did not afford the luxury of time and the diminished ranks of BR staff made the lending of staff to engage in effective transfer activities impossible. In early March, the railway close to the front lines moving up the Rangoon–Mandalay Trunk Line was understood to be under Chinese control. Chinese overt control extended from Kyaukse down to Toungoo.[31]

Beneath the surface, railway operations paralleled the command confusion that characterized military operations in this theatre. Railway operations were chaotic, consisting of a makeshift layering of different elements of British and Chinese railwaymen and British, American and Chinese military officers and men. Some of them operated under clear orders, others acted on their own initiative under orders that only hazily touched on railway operations. As far as railway operations were concerned, users were not subject to any clear chain of command. Worse, the Chinese and the Burma authorities had a different sense of ownership of railway facilities and equipment. BR men complained that Chinese officers believed that they were entitled to gather rolling stock *ad libitum* and hold it in readiness until they needed it. Even the Chinese junior officers, they complained, were used to gathering trains and supplies by force.[32]

British military and civil authorities completely pulled out of Toungoo on 20 March, handing control over to the Chinese 200th Division that had already taken up positions in the town. At the time the British were pulling out of Toungoo, the Chinese, who did not have their own railway staff there, initiated denial operations to prevent railway facilities being left intact for the Japanese. To help with the transfer, the Malerkotla Field Company had remained temporarily behind when the 1st Burma Division had moved west. These engineers were assigned the task of aiding Chinese engineers in route denial demolitions, as the 1st Burma Division withdrew, not only preparing some bridges, including the road and rail bridges at Pyu, in collaboration with Chinese engineers but also delivering explosives to the latter to undertake demolitions on their own.[33]

One of the problems with the sources for this phase of the campaign is that in addition to the issue of what is akin to under-sampling (building a broad case on isolated and perhaps unrepresentative examples) was a misunderstanding of the issues involved in the production of early reporting. The CLOs could not be everywhere, all the time on the line, and even when they were present, not all situations were clearly understood by all parties involved, in part because of limited English on the part of the Chinese and limited Chinese on the part of the British. These limitations were understood at the time and so too were the differing perspectives of military and civil officers. However, neither the army nor the railways gave sufficient weight to the prevailing confusion in accepting at face value reporting from the front lines in this particular area of the network or on either (1) the Lashio line about the same time or (2) the Myitkyina Branch in the last weeks of the campaign. The main problem on the Toungoo Front Area, where mainly civil British railway staff acted in support of mainly Chinese military forces, was actually the same problem of differing civil and military perspectives that had been the case in the Rangoon area between Biddulph and Brewitt's staff on the one hand and Milne and Chance's group, on the other. The language or cultural barrier, however, meant that the same civil–military issue was misunderstood instead as an operational culture conflict or even as a civilizational failing on the part of the Chinese. Some authorities, including Biddulph, were willing to accept at face value criticisms of Chinese military actions by BR men without considering what the underlying military motives might have been for explaining Chinese behaviour.

Among the many problems on the Toungoo Front was that Burma civil railway officers and even some military engineers did not understand why Chinese military officers were ordering the destruction of friendly rail facilities. Carr, for example, appended to Lewis's report of 19 March a series of complaints about the Chinese approach to the railway. The line, he explained, could no longer be worked down to the former railhead at Pyu because the Chinese had the night before pulled up the railway track between Pyu and Zeyawadi. For some reason, but certainly for their own purposes, Lewis continued, the Chinese had destroyed the railway facilities at Toungoo by fire. This including torching the railway quarters, the loco shed, one of the signal cabins, the railway yard buildings, St Luke's Church and so on. Lewis had informed Carr on 20 March that the line south could no longer be worked because the levers and connections in the signal cabin were warped and shunting at Toungoo south-end was no longer possible. Of what remained at Toungoo, three working engines were to be sent north to Yamethin and other engines were rendered useless to the enemy; a breakdown crane that had derailed was also to be destroyed if it was not possible to move it north.[34] Although the Chinese were criticized by railway officials for prematurely destroying the railway facilities at Toungoo events revealed that the Chinese were taking the appropriate actions in the circumstances based on an accurate assessment of the front-line situation. A railhead so close to the front was untenable. The Chinese were encircled at Toungoo on 24 March and broke out five days later permanently abandoning Toungoo to the Japanese.[35]

Failures in communication were clearly at work in the confusion over the status of railway assets at Toungoo, but what made these failures especially volatile were the draconian measures applied by the Chinese Army, which was, by this time, already in its fifth year of war with the Japanese military. As an example, in one incident around 19 March, a Chinese troop train heading south to Toungoo witnessed Burmese looting goods wagons sitting in a siding north of Thazi. The Burmese, numbering perhaps as many as 500, were local villagers. Some accounts claim they were only armed with household *dahs*, a kind of large, all-purpose knife, but other reports referred to rifles and Tommy guns as well. Two platoons of Chinese soldiers detrained and opened fire with automatic guns killing sixteen outright. The Chinese troops, following methods to prevent looting applied in China, dragged the bodies of the Burmese looters and laid them out in a row alongside the front of the looted train as a warning for other Burmese, but it must have been noticed by BR staff as well.[36]

The British assigned to the CLM the task of reducing the friction between these different operating cultures. The CLM had been set up originally for normal inter-army liaison with the Chinese forces. A British CLO was assigned to each Chinese Army, division and regiment. The GOB also assigned civil CLOs as well as a special CLM staff at Lashio. The first Chief CLO had been Maj. Gen. L. E. Dennys, the British Military Attaché to the Chinese Government in Chungking, who had been killed in an air crash on 14 March 1942 and Brig. J. C. Martin replaced him. The Chinese officers lost some respect for British CLOs as the latter wound up being forced to undertake administrative work. Chinese officers saw administration as 'unworthy of the consideration of a fighting soldier'.[37] Unfortunately, most of the British CLOs who had been assigned could not speak Chinese. They had to depend upon young Chinese students whose command of English was sometimes imperfect and on occasion caused miscommunication.[38]

A crisis in local railway leadership occurred at the beginning of the last week of March. The loss of Toungoo, which was now surrounded by the Japanese and had already been subject to denial operations, meant that the next big rail facility north was Pyinmana Junction. Unfortunately, Lewis was now put out of the picture when he was hit in a Japanese air attack on Yedashe on 25 March. He sustained a wound that required his leg to be amputated below the knee and that he be evacuated to India.[39] Absent Lewis's leadership and with reports that the Japanese were only twenty-five miles away to the south, railway staff in Pyinmana now fled.[40] The Chinese forces moving south at the end of March thus found empty stations with no supplies, congested tracks with no help from the BR and little guidance as how to manage the railways.[41]

The fact that Pyinmana Junction was not in operation worsened the overall prospects for the defence of Burma. First, the abandonment of Pyinmana brought a closure of the Kyaukpadaung Branch without any locomotives on it, rendering the entire branch useless. Biddulph, in his 'Narrative of Military Railways in Burma', written soon after the campaign, writes that this was perhaps the most serious consequence of what he called 'the Pyinmana Incident'. Although saving the junction with Pyinmana in itself was not vital to the imperial defence to the west, keeping the Kyaukpadaung–Taungdwingyi section working would have afforded significant utility in the event of a further withdrawal. However, this required locomotives along with stock on the line itself before cutting the junction began. Although Brewitt was taking steps at the time to ensure that such equipment was on the branch, this was not achieved before the branch was abandoned and clipped off the network without locomotives. Biddulph thus gave orders that BR men reconnect the branch 'at all costs' and thus restore it to network access.[42] Officers were despatched and serious efforts were made to get the branch going again but to no avail.

Second, the closure of the Kyaukpadaung Branch also exposed the railways in the Toungoo section to a fuel supply problem inherited from the pre-war dependence on imported coal from India. Coal supplies had come from India through Rangoon, Bassein and other ports that had fallen to the Japanese. Labour problems had already created temporary coal crises early in the campaign and the possibility of coal being cut off was already being discussed in early February 1942. During this period, coal was bought from other government agencies that helped the railways to tide things over until the evacuation. However, in order to achieve this, the railways had to reduce passenger services to conserve whatever coal they received. By 3 March 1942, the remaining train services from Mandalay to Thazi, Shwenyaung and Lashio were 'cut to the bone', with only one train a day moving south from Mandalay. Services had been reduced in order to conserve both coal and lubricants that were in short supply. At full capacity, coal supplies would last about fifty days, but with the reduced services, it was hoped, they would last somewhat longer. After the fall of Rangoon and the loss of other coastal ports like Bassein, maritime supplies of coal were cut off completely. With the road being constructed from India not being completed yet, no coal was coming overland either.[43]

As coal stocks declined, alternative fuels had been considered. One possible source was wood. Even before the campaign, the BR had in normal operation several older generation engines that had not been converted to coal use. When coal shortages began in 1941, before the campaign, the BR had begun to experiment with the use of wood. As the main line ran for the most part through deforested country, wood had to be cut elsewhere and railed twenty-five miles on the Kyaukpadaung Branch up to the junction. From Pyinmana Junction it was then distributed along the line. To keep larger engines in steam using wood fuel, it was necessary to attach two fuel wagons behind the tender. Workshop staff had the adequate parts available to convert some of the engines from coal to wood burning. The main problem generally with conversion to wood fuel, however, was that engines that had been built with wood fuel in mind are given a larger grate area. In the case of coal-fired engines converted to wood use, it is more difficult to keep steam production up with steam consumption. A train with such an engine fuelled by wood could only be pulled at a speed slower, and with far less pulling power, than if that engine was fuelled by coal. It was quickly discovered that while it was difficult to run large engines on wood in normal circumstances, it was impossible on steep hill gradients. More importantly for the role of the railways in the FBC, conversion to wood contributed to congestion of line traffic by slowing trains and requiring more trains to move the same loads.[44]

Despite the challenges of using wood in engines built for coal, large stocks of wood were ordered. Nevertheless, there were substantial delays in obtaining the large stocks of wood required, in part because of the demand for sleepers sent to the Yunnan-Burma Railway project, but also because of the loss of railway labour due to flight after the December bombings of Rangoon, thus forcing the railways to rely on contractors for the wood supply. Before the wood was delivered it was thought premature to begin converting most of the engines to burning wood. Nevertheless, by 13 March, desperate railwaymen were moving up and down the line to gather wood to use for fuel. As coal supplies were cut off, the railway staff, by this time having relocated from Insein to Myitnge, also experimented with converting their powerful, hill-climbing Mallet and Beyer-Garratt engines to burn oil while heavy goods engines were now converted to burn wood. The residual oil from the Burma fields was too expensive for engine fuel and railway authorities were not offered Burmese oil for engine fuel until the oil refineries at Syriam were on the verge of being lost. However, although indents had gone to India repeatedly for the fittings necessary to convert the fireboxes to burn oil, these fittings never arrived.[45] With access to the Kyaukpadaung Branch and with it Taungdwingyi and access to Yenangyaung, the railways would be dependent on residual stocks of fuel at the stations, the latter in the meantime being bombed repeatedly by the Japanese. Had railway activity continued into May in this section, the acquisition of additional supplies of fuel would have become a pressing issue.

Coward at Pyinmana

In addition to the already onerous assortment of difficulties in keeping the railway line open was the rapid arrival of a huge volume of new users of the railway. Chinese forces began moving south to take over the front from the British. There were few railwaymen on the line, however, to operate the line in their support. Chief CLO Martin thus telephoned orders to Mandalay to have senior railway staff experienced in traffic control to aid the Chinese. By coincidence, Coward was already on his way on a southbound train, despatched by Brewitt to undertake a survey of the situation of the railway and a census of the staff and stock in the Front Area. When Coward reached Thazi on 27 March, en route to Myingyan, he was informed that Pyinmana had been completely abandoned by BR staff and was out of commission and that Chinese troops were moving south to Yedashe and someone had to open the junction. Telephonic communications were unstable, but Coward was able to order George Peters to return to Pyinmana and get the junction there working until Coward could arrive, while he was also able to telephone Brewitt who sent a staff car to pick Coward up and race him to the junction. The timing was crucial. It revealed the fragility of railway operations outside of the major yards at Rangoon and Mandalay that made fast movements and decisive action essential. While Coward was en route to Pyinmana and six Chinese troop trains were barrelling down towards the Pyinmana Junction with a half dozen railway staff under Peters at that junction, the Japanese launched a heavy bombing raid on the railway yards there, closing the line. Coward, accompanied by Major Snow, CLO of the 22nd Chinese Division, retrieved a special engineering train known as an 'accident special', and reopened the line the following morning. By midnight they were able to get all six trains of Chinese troops through.[46]

Coward's main task would remain the running of trains on the section of the line between Advanced Chinese HQ at Yedashe north to Advanced Army GHQ at Pyawbwe. These railway operations on this lengthy section were just barely held together by Coward and four subordinates with mostly non-existent communications with Brewitt's HQ. Japanese bombing constantly brought communications down and made even existing communications difficult. Coward would have to run from one section of the yard to another to reach the only telephone line that remained intact. Without sufficient staff, he was often not free to do so and in his absence telephones were additionally looted. Air raids had inflicted more serious damage. Only three lines were functional, the rest having been knocked out by air attack. Both of the cabins in the Pyinmana

Railway Yard had been knocked out. Although the railwaymen had to work in darkness, there were also no lamps of any kind in the railway yards, either for shunting or for lighting any part of the station buildings. The Stationmaster's controls were also out of commission.[47]

Whether and, if so, how Coward could continue running railway operations in this part of the network was unclear. There were no staff on the Taungdwingyi Branch or at any roadside station between Yedashe and Kyaukse Although Coward's staff were in favour of abandoning Pyinmana altogether, Coward considered that with no other railway staff in the Forward Area he would make Pyinmana his HQ. Additional help was on its way. On 28 March, Martin, General Wong and other officers then went down to Pyinmana to arrange for the movement of rations down the railway line to Chinese troops at the front lines. According to Martin's report, Chinese officers had begun to bully railway staff and most of the railway staff ran off, leaving behind only the stationmaster and two engine drivers. The stationmaster agreed to do everything in his power to get his staff back together. The report warned that if better treatment by the Chinese Army of BR staff was not ensured, the native staff would abscond and the Chinese would have to run the railways themselves.[48] Martin's report has to be understood in the appropriate context, however, as this sector of the railways ran all the way up to Chinese positions on the front lines and 'native staff', presumably Indian, were already unnerved by the constant and immediate danger of being killed. Indeed, just five days later, on 3 April, when the Japanese again bombed both Yamethin and Pyinmana stations, the native staff again threatened to run away.

It was during this period that some of the experiences that would shape how the railways characterized the relationship with the Chinese were first circulated. Coward was the source of many of the complaints about Chinese abuse of the railway that worked their way up the chain of reporting. Coward was the source of the complaint that Chinese were firing on railwaymen at work in the yard. He was also the source of the story that a Chinese officer had accused a railwayman of being a spy because the railwayman was shunting the engine for reasons that were unclear to the Chinese officer.[49] Arrangements for security for the line became necessary. On 31 March, General Tu agreed to the suggestion by a CLO to post a guard of fifteen men under a major at the Thazi, Yamethin and Pyinmana stations to prevent Chinese troops from interfering with railway staff again. The following day, when problems continued, General Lin also issued orders to Chinese troops not to interfere with Burmese or other railway staff. The same day it was also promised that the sending of Chinese railway operating

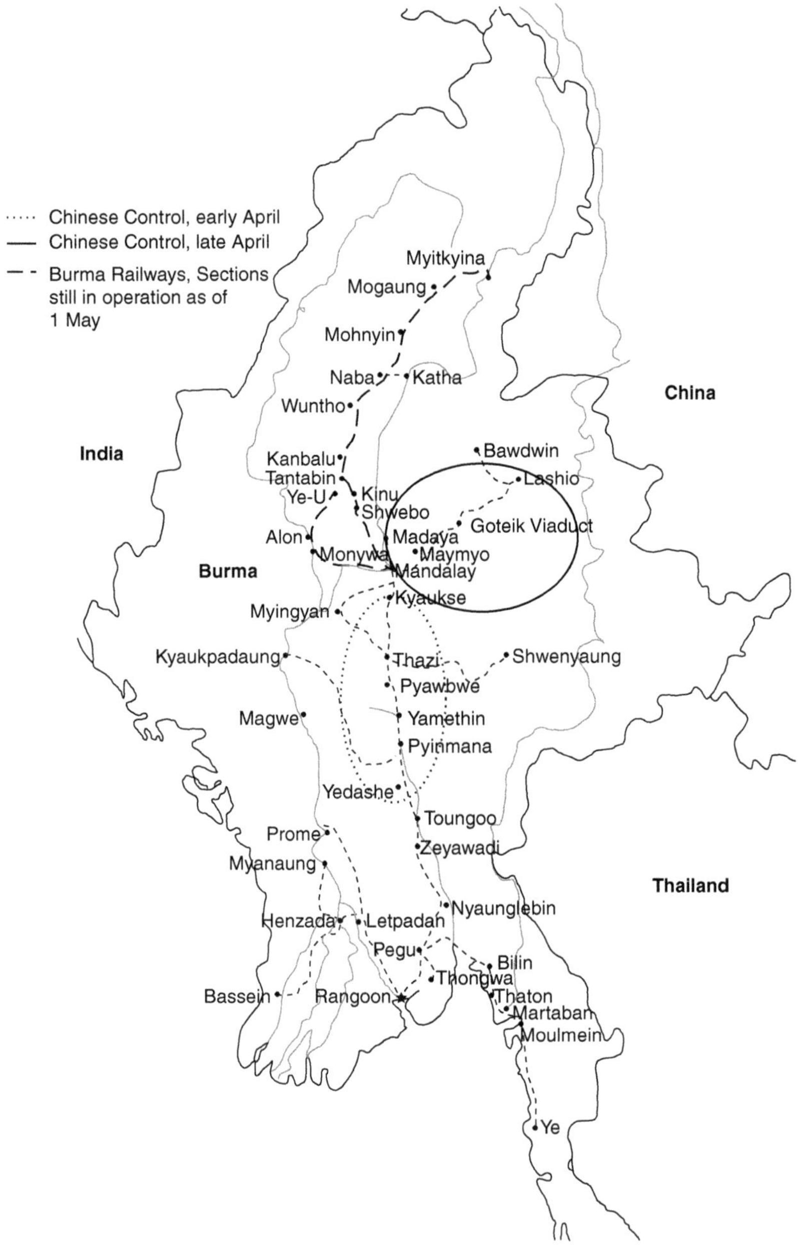

Map 5.1 Burma Railways in April 1942.

personnel would be hurried. On 1 April, General Tu also agreed to supply a guard of twenty men, on trains running south of Yamethin, for track maintenance. Coward agreed to run a regular supply train between Pyinmana and Yedashe, trolleys carrying petrol between Yamethin and Pyinmana and making available on a daily basis sufficient rolling stock to evacuate 200–300 men at Pyawbwe.[50]

The most difficult challenge was keeping up sufficient numbers of staff for working the line. On 1 April, the morning after the railway crash at Pyinmana, Coward found that Peters and the rest of the Pyinmana staff on the railway had left, boarding an evacuation special running north. Coward located the RTO with a car and the two raced north to outrace the train, successfully heading off the special at Yamethin. The same night, Coward with the help of Lieutenant Hosie and several additional subordinates provided by Officer Commanding, Railway Transport, took Peters and some of the staff back to Pyinmana. The Loco staff, however, had successfully mingled with refugees and could not be located. Fortunately, some staff had remained in place, such as Mr Merritt who had kept the Pyinmana boiler house going. Coward also adopted some measures to ameliorate the dangers of working in the yard. The staff would no longer be required to work in the daylight, only at night. This, Coward lamented, 'was the only way to keep the staff from absconding'.[51] Military engineering help was also lent at Pyinmana at the beginning of April in the form of Carr and his 139 ROC, IE. Carr and his men would work as drivers and as firemen and worked all the trains in the Forward Area where they were continuously employed.[52]

Coward complained after the campaign of the 'almost complete absence' of communication with Brewitt's HQ. As a result, Coward found himself working entirely on his own initiative a long stretch of line that included two railway yards, one at Yamethin and at the other at Pyinmana, both of which had been bombed heavily and from which the railway staff had absconded.[53] Coward's measures were sufficient to keep the lines working down from Yamethin until another Japanese bombing raid completely levelled Yamethin. Coward explains that the loss of Yamethin added greatly to their difficulties in keeping the railways running on the section, for now all of the Yamethin Loco and Traffic staff ran away. Coward thus split his ten remaining men into two groups, posting Lieutenant Hosie and five traffic subordinate officers and one British, Other Ranks (BOR) at Yamethin and kept the remaining four traffic subordinates and an AFO at Pyinmana. A day or two later, Hosie's men except for the BOR had also absconded. As a result, Coward had to work two yards, the one at Yamethin and the other at Pyinmana, with a total of only five men as his staff, personally accompanying trains up and down the section. Additionally, as the bombing had damaged Yamethin's water supply, the engines had to be watered (just as they had to be coaled) by hand, exacerbating the shortage of labour. Coward took to 'persuading' refugees to help water the engines. Communications had to be maintained across the section from Coward's HQ at Pyinmana by Coward running across the yard to the Locomotive coaling hut north of the bridge, where the phone line was still intact, for all the posts between the hut and the main station were down. Brewitt joined him several days later on 5 April.[54]

Some help came in running traffic in the Chinese-controlled section of the line from the Chinese Army in the second week of April. On 9 April, a Chinese railwayman, Brigadier Tseng, arrived at Yamethin with a group of other Chinese railwaymen, including a number of drivers, to undertake a significant portion of the daily operation of railway traffic in the sector. A group of nine of these Chinese drivers under a Mr Hu would arrive at Thazi a week later on 16 April. Tseng also requested even more Chinese engine drivers be sent from China but their arrival was delayed until the last weeks of the campaign and although they did actively take over the Lashio Branch briefly in late April, they were too late to help resolve problems in the Chinese sector on what had been the Rangoon to Mandalay Line. The Chinese railwaymen were able to help relieve some of the burden on the BR staff undertaking a large share of the work on the line in the Chinese sector.[55] Nevertheless, Tseng's arrival was greeted with the same negative reaction that had greeted other outsiders to BR operations thus far in the campaign. Again, the Chinese railwaymen were bound to be a problem because they 'did not know the road'.[56] British reports complained that Tseng did not comprehend the situation the railways were in and 'made all sorts of demands for trains and movement of troops which any clear thinking railway official would have immediately realized were impossible'.[57] But there were also other problems that limited their already more circumspect role as drivers. For one thing, they were not familiar with the engine types being used. BR staff also complained that Chinese officers forced them to break rules and regulations.[58] For the Chinese officers, the BR's rules of operation seemed be bewildering and unnecessary arcana. Stories of clashes on the Toungoo section of the line circulated everywhere among railway staff on the rail network.[59] Complaints even began coming into the Transportation Directorate about Chinese soldiers threatening to shoot station staff when trains were held up because they ran out of coal or water or had to wait for the passage of another train to access the main line. As Milne recalled, at Myohaung, Chinese troops threatened to shoot a Burmese Assistant Traffic Superintendent if he did not let their train go through.[60]

The clash of different combat cultures and the relationship of the railway to these respective cultures was partly responsible for these problems. As Col. James Vivian Davidson-Houston observed, 'the difference between their ideas on rail movement frequently led to misunderstanding and difficulties'.[61] The Chinese divisions had just come from a country that had already been at war for five years and the organization of its forces was made in a situation where they were fighting on domestic soil and could thus live off the land. Chinese armies were thus not supplied with administrative or supply services. Local civil authorities were responsible for supplying rations and transport for troops. All the resources of the locality were at the disposal of the Chinese forces. This had in fact been one of the main reasons that the British had earlier refused Chiang Kai-shek's offer of Chinese armies to help defend Burma as without logistical support these armies would have overburdened existing transport resources in Burma although at the time they also sought to demonstrate that Britain could handle imperial defence on its own.[62] This combat culture was very different from what the civilians of the BR staff, now in uniform, and even the Indian Army were used to. This also included summary justice in response to the breakdown of civil order (although the imperial forces were also very close to this point).

Pyinmana would soon after be obliterated as a railway installation for the remainder of the campaign. On 15 April, the Chinese forces desperately tried to save their divisional artillery by moving it northwards through Pyinmana. Given the importance of the artillery, they insisted that its passage could not wait for night and that the railway staff would have to work the yard in the daylight. Unfortunately, the Japanese knew that the artillery was coming through and thus caught the railway staff engaged with a train in the yard containing both artillery and ammunition at 1600 hours. The heavy bombing that ensued was a 'plastering' destroying nearly the entire railway facility.[63] The railway operations by this time were being overwhelmed by wounded from the streets and thousands of refugees moving through every day, requiring ambulance trains and evacuation specials. Further, the 15 April air attack had made a direct hit on the Pyinmana tank and boiler house and so there was no water now on most of the section except for the handpump at Pyawbwe. It was estimated that it would take at least three days' work to put the yard and the pump house back into operation at the time. Only one line remained suitable for the passage of trains.[64]

Coward now moved his HQ back up the line to Tatkon while the Chinese 22nd Division withdrew back fifteen miles to the north to Yezin. Although the HQ of the 96th Chinese Division remained at Pyinmana, the latter was no longer viable as a junction and it was too close to the front lines (and indeed, the Japanese would occupy it on 18 April). The Japanese were now sending low-flying fighters to machinegun the line between Pyinmana and Yamethin. Tatkon, which was midway between Pyinmana and Yamethin, was somewhat further away, and was the railway station nearest to Pyinmana on the trunk road. This allowed him to maintain communications with the 96th Division by road. All civil CLOs had also pulled back to Tatkon. As Coward's staff moved through Yedashe on its way to Tatkon, the Japanese bombed the station damaging all the lines and killing the Assistant Stationmaster, so Coward and his men improvised and got the loop functioning to a degree. Ywadaw, another station on the line, had already been heavily bombed on 8 or 9 April and remained unserviceable.[65] At Tatkon, Coward was closely supported by Carr and a staff of about three traffic and nineteen EIR Loco men of the 139 ROC, IE, which was still insufficient for operating requirements as well as for keeping the railway in repair in a Front Area with the Japanese bombing and shelling. Coward's biggest complaints, however, were the lack of ration provision for his men, the lack of support provided by the colonial Engineering Department and security in the area.[66]

The situation rapidly deteriorated from this point. Most work at this point consisted of downwards despatch of arms, rations and other supplies. Coward hoped this would only be so far south as Yedashe but as it was too far from the trunk road, he had to make Kyidaunggan his railhead for a short time. Coward found that it was only possible to work at Kyidaunggan at night, so he and his staff would remain in Tatkon during the day and then travel down to Kyidaunggan as evening approached to work in the dark. This situation lasted for a few days until Coward and his staff went to Kyidaunggan for a final time and took back all of their supplies and withdrew his line staff to Tatkon. These men were then redistributed to stations to the north from whence staff had earlier absconded.[67]

Civil disorder

There was certainly evidence of the robbing and murder of Indians on their way out of Burma in 1942 and of looting by Burmese and others of abandoned homes and railway wagons. However, the evidence for sabotage of the railway lines during the campaign by Burmese is actually very weak and was more the product of hysteria brought on by the unravelling of the colonial order in the face of an approaching enemy that was at work. However, the form given these fears was also a legacy of the same colonial order. The wartime fears of Burmese fifth columnists built upon decades of pre-war, colonial distrust of the Burmese as saboteurs, an image in part that helped to mask the inadequacies of colonial investment in infrastructure in an effort to save costs and increase profits. They would also play an important role in shaping the perspectives of others in the campaign and their testimony. For our purposes here, the most important outcome was that Coward's descriptions of what he and his railwaymen in the Pyinmana area termed dacoity would contribute to the historical narratives of the campaign and the railways' place in it.

It was no coincidence that the main official pre-war investigation of railway sabotage was initiated after and focused upon an incident on what was now Coward's section of the line. In 1940, a number of derailments occurred in one particular section of the Rangoon–Mandalay Trunk Line. There was a derailment on the bridge between Penwegon and Tawgywe-In on the first occasion and a derailment between Daiku and Pyuntaza on the second occasion. The latter accident killed five and injured twenty-five, creating a public outcry.[68] In response, a high-profile committee (the Swithinbank Committee) was appointed by Prime Minister U Saw in 1941 – including among its members a Burmese senator, U Kyaw Zan; a former mayor of Rangoon, U Ba Gale; the Burmese Chamber of Commerce representative to the Burma Legislature, U Aye Maung; with B. Swithinbank as its chairman – to investigate. The Swithinbank Committee examined the police case files of a number of derailments that spanned the course of twenty years, believing them to look connected, but finding no common motive, nor any evidence to connect them. In these cases, fishplates were removed and rails slewed near a bridge or jungle to inflict greater damage in the first case or an easier escape under cover in the latter case. The motives were particular and different, for in one case the saboteur wanted to get someone into trouble, in another case, someone was trying to obtain loot, and in yet another case, the motive was political. Ultimately, several Burmese were executed because of police efforts. The committee submitted its report to the government in 1942, only a few weeks before the evacuation of Rangoon and shortly before all of the railway derailments in the north, far away from the scene of the pre-war derailments that had been investigated. It would seem very likely that the tension and fear of the time led to hysteria that produced and further fed into the belief that antinomian villagers were at fault for any railway accident caused, regardless of the absence of any visible evidence as to their culpability. Further reinforcement of these suspicions came from the rapid occurrence in a short space of many derailments and the publicity of a government report just a short time before which implicated Burmese villagers as responsible for railway derailments.[69]

The Swithinbank Committee report also directed an angry finger at Chance's Railway Board. It seemed uninterested in dealing with the cause of derailments and suggested that this was work for the police. The Railway Board was also uninterested in adopting mechanical measures to make derailments difficult. Chance's administration, however, rejected such measures because they would be too expensive. The Swithinbank Committee thus threw up its hands and suggested in the face of an intransigent and disinterested Railway Board that all they could recommend were that (1) patrolling on the worst section of the line should be increased, (2) the jungle should be cleared where derailments occurred, (3) the police should be proactive in dealing with criminals in these areas, (4) Burmans should replace Indians as Permanent Way Staff and (5) that the Superintendent of Railway Police should attend or send a representative to departmental enquiries which were made by Railway Police into charges that were made against railway employees. There was no time for such measures to be undertaken, but given the Chance administration's attitude towards improving security, it is doubtful if these measures would have been adopted even in the best of circumstances. The measure that the Railway Board did undertake was to instruct the Traffic Department to run a pilot engine in front of mail and express trains at night, so that they would hit any breaks in the line first and thus give warning to the train behind it.[70]

Fears of Burmese sabotage of the line surfaced again in March and April 1942 in the Pyinmana area. Coward advocated to railway authorities the necessity of patrols on the line not only to protect railway sections but also to assist in uplifting the morale of staff, those at roadside stations in particular, by helping with the running of trains, providing communication and guarding railway staff while they worked.[71] He was motivated not just by the sagging morale of the railwaymen who kept the railway going in this sector, but also by fears of Burmese assault on personnel and sabotage of the line. These fears emerged and became primary among BR men to a degree that was peculiarly colonial, for civil railwaymen in Europe did not fear the depredations of the civil populace upon them when conditions worsened, but they became terrified of this eventuality among a colonized and subjugated population whose loyalties they doubted. As a result, in Rangoon, on this section, and then again on the Myitkyina Branch at the end of the campaign, this fear came to the forefront of reporting on the conditions among the railwaymen working the line. Coward, for example, complained extensively about the fears of the railwaymen under him of dacoit activity. As Coward related, railway staff at roadside stations on this section were 'even more terrified of dacoits than of bombing and machine gunning'.[72] Nearly every station up to Myohaung, he warned, had been looted and 'large bands of dacoits and armed deserters [were] roaming around as they chose'.[73]

Coward's wartime reports of sabotage in his section of the line, which found their way into anonymous references in more generalized reports, spread like wildfire in a colony at war, when it was difficult for many authorities or the European population to suspect derailments as anything but the result of Burmese sabotage regardless of the actual causes. Correspondents such as Leland Stowe related the warnings they had received, presumably from white colonial residents, never to show that one had money for the Burmese were good with a knife. Stowe believed the problem was racial stratification in the colony and the resulting distrust between ruler and ruled led him to believe that the colony was full of real and potential agents for the Japanese invaders.[74] Coward's experiences on this section of the line, however accurate but peculiar they may have been to this area of the railway network, lent confirmation to these fears.

Perceiving the railway derailments as purposeful sabotage only added to the atmosphere of fear and panic. Certainly, some sabotage by the Burmese thus took place, or even increasingly took place, on the kind of mundane scale that was possible for villagers who had no direct experience in railway management. In early April, Coward's staff reported specific problems that they suspected were sabotage. These included the removal of several nails and fishplates on the Pyinmana to Taungdwingyi section of the Kyaukpadaung Branch.[75] These local railwaymen's perspectives shaped those of newcomers to this section of the line. American officers who arrived on the line to help the British with railway support for the Chinese, for example, quickly attributed the opened switches, wrecked signal boxes, loosened nails, and freight cars dragged across the lines they encountered on Burmese artifices directed at keeping Chinese troop trains from running to the front lines. As a result, one of these officers went up and down the rails in a special rail car, Tommy gun in his lap, 'inspecting every foot of the way for sabotage'.[76]

As the front lines moved north and the railway compacted in on itself, accidents and accusations of sabotage increased rapidly. However, these reports increased as a climate of fear ensued and intensified and as the active maintenance of the tracks and the signalling equipment came to a halt in most areas. Despite their robust signature on the landscape, railway equipment and the permanent way were highly sensitive to climate, erosion and wear and needed constant care. By mid-April 1942, railway repairs were extremely ramshackle and trains were often being organized and driven by men who had either never been engaged in these particular tasks or had been away from them for many years. Everyone's nerves by this point must also have been on end.[77]

Derailments during the late stages of the FBC occurred in a colony where the BR Board had already responded in 1940 to the increasing costs of railway materials brought by the outbreak of the war in Europe, to simply delay essential repairs for as long as possible.[78] It is probable that some crashes in 1942, after the war with Japan had broken out, were caused by human error in a difficult, desperate situation where fatigue, lack of repairs, overcrowding and unsafe railway operation were regular conditions, making worse a railway network already in disrepair. In fact, blame for the various acts of sabotage identified on the railway network was directed at the loss by the railway of much of its labour, such as the permanent way gangs who could otherwise have patrolled the line and thus have prevented purposeful damage. Their absence would also suggest, however, that the wear and tear that had to be fixed on a regular basis in peacetime under normal loads was no longer being repaired.[79] One driver recounted very soon after his escape and arrival in India that 'slight mishaps' had begun to occur in April simply because the strain was taking its toll on the over-worked men, rolling stock and locomotives.[80]

The means to quickly remove derailed trains or wrecked equipment was also more difficult to come by in the war, and other wartime conditions hindered the quality of repair work done. During normal times in India and Burma, breakdown or accident trains were kept on hand at the important railway stations and junctions. Such trains could not be removed quickly without running over a number of sets of points before getting onto the main line. This made them vulnerable to damage or destruction when main stations were bombed. After the loss of Rangoon, the BR had a number of breakdown cranes, including, according to tonnes of capacity, eight thirty-tonne steam-powered cranes, two twenty-tonne hand cranes and many smaller five- to twenty-tonne traffic hand cranes. However, these were scattered around the network and as congestion set in and demands overwhelmed labour, it was difficult to apply them rapidly when needed, or at all. The establishment of a special engineering train that was always on the tracks aided keeping the railway lines in repair despite repeated bombing. This was placed under the command of a railway officer and staff to supervise repair work. It carried labour, equipment, stores and even rations. However, in the last month of the campaign, even this option was no longer available. Now, as repairs to the line and clearance of wreckage could only be accomplished quickly, often with untrained labour, the railway line itself became increasingly dangerous. Repairs along the line, for example, were on occasion made over bomb craters and these were often quickly or recently done and unstable.[81] Caution thus had to be continually exercised, for there was probably little else that could have been done in the circumstances to rectify the problems that now derailed trains up and down the line.

After the destruction of Mandalay, the great eructation of administrative staff of all kinds out of the city, including the railway personnel, created what seemed to be ripe opportunities for disorder, robbery and violence as it would in most societies where government was collapsing.[82] The outbreak of real acts of violence and crime would add to existing anxieties among the railwaymen and others over the perceived readiness of Burmese villagers to sabotage the line. While sabotage was possibly at work in some cases, in particular in Coward's sector where anxieties over Chinese military excesses angered local Burmese, there was often only circumstantial evidence to suggest that it was. Indeed, despite the railwaymen's fears, most Burmese villagers then and during the Japanese occupation were generally content to remain merely spectators to the whirlwind of events that they could not control and which, no matter how things went, did not seem at the time likely to change their lives. Indeed, the general apathy of the 'Burman of the plains' was one of the chief factors that Slim considered as working against his plans during the campaign. This was partly to blame, he felt, on a colonial administration that saw folly in doing anything that would encourage despondency or alarm among the general population and thus had done nothing to mobilize them against the Japanese, not even a Burmese Home Guard or a guide or civil transport organization.[83] Colonial rule and the railways had given Burmese little evidence that what was happening between the British and the Japanese had very much to do with them.

Station-level leadership

One of the key factors in maintaining staff on the line during these weeks was station-level leadership. The work demanded of the staff by this time in the campaign regularly seemed to be surpassing the limits of normal human endurance. They were also concerned around the clock about the approaching

Japanese Army, the threat of bombing and, however exaggerated, sabotage or murder at the hands of the local population in the context of the collapsing colonial order. Nevertheless, a good stationmaster on the platform and foreman in the yard could keep the local staff together and working. A good example of this was Thazi Junction in the first two weeks of April. During this time Thazi Junction had undergone two heavy bombing raids. Although the damage to the railway area was not very extensive and the loco shed and railway yard remained relatively unscathed, the raids were enough to frighten staff, but most remained at their posts in no small part due to the fortitude and local leadership at the station. For unclear reasons, the foreman, Mr Hill, was granted leave after the bombings and without a foreman the railway staff had succumbed to rumour and the climate of panic as refugees fled north and fled themselves. As a result, by 13 April, Thazi Junction had only four railway staff left. Railway authorities quickly sent a new officer to reform the staff. With new leadership in place, old and new staff were mustered together, with about thirty–forty staff by 15 April, some staff drifting back in and others being sent by Coward. The junction was running again in a few days. The lesson drawn by the railway authorities was that had local station leadership remained intact, that is, had the foreman not been granted leave so close to the Front Area, many more staff would have remained at their posts.[84]

The need to maintain staff in place for railway operations was crucial because of their technical requirements and the balance that needed to be maintained between different responsibilities to keep locomotives running and the railway functioning and in good repair. During the next week, even with two score of men, Thazi Junction found it difficult to work the line in the Front Area. Six men working as boilermakers and as fitters were insufficient to keep engines in the sector in running condition. Only limited and essential repairs could be undertaken, including replacing worn brake blocks, cleaning and replacing injector cones and grinding in valves. Washouts by contrast could not be done. The shortage in running staff was even more severe, although this could be more easily met, at some risk, with military men on loan. Bona fide engine drivers were forced to work long hours with little rest. Their labour was supplemented with inexperienced men drawn from the Burma Rifles who acted as firemen. There was no means of keeping track of how many hours each man was working on the section. The recommendations were that, in future, drivers carry around a card on which foremen could log departure and arrival times. Fuel coolies were also no longer available and coaling could only be done by using military men or asking railway staff to double as coaling staff. At Thazi, they mobilized ten men trained by the RTO, Captain Williams, to help work in the shed as coaling staff or as 2nd firemen, and carriage and wagon staff were also asked to coal engines in addition to their normal tasks.[85]

By mid-April, Thazi Station staff were only held together by the leadership of the RTO Captain Williams with the support of the stationmaster and the assistant stationmaster Hussain. Thazi had 'plenty of engines' but difficulty in locating key staff and staff complained of a lack of pay or rations. Thazi no longer had a Loco Foreman, although Williams was able to muster a total staff of three drivers, two firemen, one first fireman, two cabinmen, one brakesman and one porter. Williams ordered Indian soldiers to act as firemen as well and used the brakesmen to work as a number taker and sent in an urgent request that a Loco Officer be posted at Thazi as soon as possible. The main complaint by the stationmaster at Thazi was that he was being 'badly interfered with by Chinese troops'.[86] Williams was able to get the station in sufficient shape to keep the Shwenyaung Branch working as far as available staff levels could permit. The same could not be said of the moribund southern Myingyan Branch. By mid-April, Meiktila Station had no staff at all, no RTO could be located in the vicinity and although there were many wagons in the yard, many were 'very badly damaged'. Although the staff at Myingyan Station continued to work normally, no trains were running any longer on the section between Myingyan and Meiktila.[87] The branch would remain dead for railway traffic for the remainder of the FBC.

While railway staff at Tatkon and Thazi kept operations going as best as they could, the staff at Yamethin was even more threadbare. Lieutenant Hosie maintained Yamethin with only three men, including two signallers and one guard (named Ramphair) who had no rations and worked in the yard every night. There were no traffic staff, the Yamethin pump house had been bombed and it was impossible to water any engines. In addition to four empties in the yard, there were twenty loaded wagons whose contents remained unknown, as there were no staff to check, although Hosie soon received orders to do this himself to see what was inside. Although he requested six subordinates and four menials,[88] it was difficult to find a concentration of surplus staff anywhere sufficient to shore up any station. The stationmaster of Pyawbwe, Maung Than Pe, also absconded, abandoning the station by mid-April, and left the two assistant station managers to fend for himself or herself. This included Maung Ba Win, who took charge and Rampal Singh who assisted, supported by a staff of one CBC, one gateman, two porters, two jamadars and one sweeper. Again, the lack of rations and pay was a major complaint of the staff.[89]

During this period, Coward again had continued to operate on his own initiative because of three main factors. First, communications were nearly non-existent and there was no possibility of operating as part of any chain of command, civil or military. The communications blackout was only partly accidental. As a rule, Coward complained, Chinese divisional HQ gave the railway no information with which to work, perhaps because of the constant suspicion that there were spies among the railway staff. In addition, by mid-April he had no wire or phone communication with Chinese GHQ either.[90] Without communications, the rumours that circulated could not be confirmed or denied and thus began to be taken seriously by remaining staff on the railway line. One such rumour was that the Japanese were making a move around Pyinmana through Lewe and were about to reach Tatkon and another was that the Japanese were coming towards Meiktila from Taungdwingyi. As a result, on 20 April, the District Commissioner of Yamethin, H. C. Baker, held a conference with Coward at Pyawbwe and Baker decided that he would leave Yamethin. Baker advised Coward to abandon Tatkon as soon as he could. As he had to operate on his own instincts, Coward decided to withdraw as well back to Yamethin. Coward found that Pyawbwe had been heavily bombed and there was no longer any water even from handpumps there. Coward thus had to take pumping staff to Tatkon daily to get water to run the engines, but only for a few days as he soon received orders that all staff were to withdraw. After immobilizing the engines at Yamethin, he went to Thazi and had the orders reconfirmed. He then handed the railways up to Thazi over to Chinese control.[91]

The second factor was that the status of railway facilities and running equipment shifted so dramatically and frequently that decisions could only be made on the basis of timely, on-the-spot knowledge only available to a local manager. This situation was increasingly due to the depredations of the Imperial Japanese Army Air Force, which heavily bombed Thazi on 22 and again on 23 April. The air raid on 22 April did only minor damage to the Loco Yard, although shrapnel damaged the tanks of three or four engines, the delivery pipe to the shed water tanks broke and the overhead tank was pierced. However, this raid did much damage to the station yard and to Chinese troop trains and supply trains in the station. The raids on the following day were far worse, with the Japanese making six to seven low-level bombing raids of the railway area. Only one line in the station yard now remained undamaged. The water main from the south end tanks broke. All lines in the railway yard except for two were badly damaged. The line from the loco shed to the station yard was so badly damaged that no engines could be taken out. The loco shed was out of commission because of blast damage. A number of water columns were damaged. The overhead tanks were again pierced in several places and the main delivery pipe to these tanks broke close to the shed exit points. The turntables were also badly damaged. The raid also represented a huge loss of railway engines at a time when they were probably needed most for Burma's defenders and the evacuation. A railway accident a few days before had blocked the line at Samon, so a train with multiple engines that had been made up to go to Myitnge on 21 April never left and was caught in the station on 23 April. Shrapnel damaged these and all other engines in the shed and in the yard, piercing tanks in multiple places, the steam pipes and even the firebox. There was now no possibility of getting them out before the Japanese arrival and so their water gauge glasses were smashed, their injector and delivery cones and caps were removed, and, while the staff were able to remove the connecting rods on some engines, they ran out of time to do so on others.[92]

The third factor was that labour was only held in place by local leadership in a situation in which only a local manager could grasp the status of staff. No reserves or replacements were being sent back south and the men had to be continually shifted around to make up for absconders, the wounded and the dead. The last group of staff remaining at Thazi, including Carr, the foreman and assistant foreman and six drivers, were withdrawn on the morning of 24 April. They used a train that had been marshalled the night of 22 April after the first raid and sent northwards when the raids came on 23 April (just in time to avoid destruction). This evacuation was the end of BR men's activities on this section of the line. The staff and equipment from Thazi were brought up to Mandalay area to support railway operations there as long as they could. Carr and his men also brought with their train the accident train, the three-tonne steam crane and wagons carrying stores and other fittings in a convoy that moved north to Myitnge, pulling the material out in time to save it from Japanese bombs and strafing.[93] They would be put to use in the railway yards at Myitnge, Myohaung and Paleik for the remainder of the fourth week of April, all three facilities now placed under Coward.[94]

On 25 April 1942, Stilwell and Alexander met at Kyaukse and decided that holding the Mandalay–Irrawaddy Line was no longer possible and that the east bank of the Irrawaddy would have to be evacuated.[95] To the south, Chinese railwaymen were still insufficient in number and in their familiarity with the network to make realistic use of it in support of Chinese forces. Even had they decided otherwise, the Chinese would have had to do so with the railway only able to provide decreasingly effective Forward Support. Nevertheless, Stilwell looked to the railways to provide one last service on the Toungoo Front, which was to aid Chinese troops on their desperate effort to avoid capture or destruction at the hands of the Japanese Imperial Army. To implement the withdrawal, Stilwell relied on Col. Paul Leroy Jones. American military officers had been operating at Chungking as the 'American Military Mission to China' but from early March as the 'American Forces in China, Burma, and India', about the same time as Stilwell arrived to take command. Stillwell had informally appointed Jones as his 'railroad operations officer', reportedly leading foreign correspondents to dub Jones 'Casey Jones'.[96] Jones used a special motor rail car to go up and down the railway line to arrange trains to move south to Thazi to help evacuate the 22nd Division. Jones recalled that he had to use brute force with the Chinese troops who did not like or trust the trains. Stilwell lent authority to Jones's orders and accompanied him personally as Jones attempted to mobilize rail transport. At one, otherwise abandoned, station, Stilwell and Jones got out of their jeep and Stilwell, approaching a lit freight car in the railway yard, harangued a railway official whom they found inside for not doing his part. Stilwell stressed that war was serious business and everyone had to forget that they were 'generals and colonels and pitch in', and there had better be some trains down to the front lines fast to get the Chinese soldiers there out. As Jones lacked railway personnel, he also had to rely on General Sung's help to get Chinese troops to push derailed freight cars off the tracks so the evacuation trains could get through. Jones spent three days and three nights at work on this withdrawal with little or no food or sleep, in one case driving a locomotive for several hours against strong rain. Jones was able to get out three trainloads of Chinese troops.[97]

The speed of the retreat and the constant danger of Japanese envelopment made any withdrawal of the Chinese divisions complicated. CLOs had also held a meeting on 26 April and decided to evacuate, believing that the Chinese forces were in such a state of panic and order so broken down that there was little liaison that could be achieved anymore.[98] When a few trains were eventually sent south by night on 27 April to one of the stations north of Wundwin in order to bring back Chinese troops there away from the front, the Chinese colonel in command asked for British troops to be posted on the trains and around the station. His fear was that otherwise the first of the Chinese troops to reach the trains would seize the trains in their panic and then force the trains to take them away, abandoning their comrades. This fear was taken seriously and the 17th Indian Division posted a small detachment of soldiers in and around the railway station. When Stilwell's staff saw the troops from the 17th Indian Division posted in the station, however, they believed them to be the tail end of the British rearguard and that the British, despite promises made by Alexander, were in fact attempting to flee north ahead of the Chinese. According to Slim, Stilwell even complained 'emotionally' to him that the latter had failed to conduct an effective rearguard action.[99] Meanwhile, a juggernaut of retreating Chinese forces threatened to overwhelm the railway up to the Ava Bridge and the entire Myitkyina Branch beyond.

Chapter conclusion

Different factors play greater and lesser roles in explaining railway success in different locations. In this section of the railway network, it was operational autonomy, or the ability of local managers to operate the railways autonomously of central control that was the biggest factor. Biddulph made this decentralization of authority in part by removing the CRC from control of the railways on 24 February, as discussed in a previous chapter, and allowing local railway managers considerable autonomy despite the subsequent migration of the civilian railway administration to the north.

Limited and unstable communications on the line during the invasion eliminated the possibility of regular management above the local manager, such as the stationmaster or Loco Foreman. Only special, personal management was possible, where a local railwayman used his own initiative to go up and down the line and hold things together with the force of his own personality and the management skill set he possessed. By all accounts, the section of the railway examined in this chapter remained autonomous of central railway authorities during these months. Although the BR staff had been militarized, the military organization provided was more symbolic than operational. The railway staff found themselves running trains on their sections as autonomous ROCs. Local authority and local responses were effective and perhaps the deciding factor in the continuity of railway operations after 24 February.

Why should decentralization have worked? Local managers 'knew the road' as BR men often said of those who possessed a detailed grasp of local operations, facilities and the permanent way. Subordinate staff could only be shifted and redistributed where they could be found and even caught on occasion for application on the section where they were needed most by men who had an intimate knowledge of the local section of the railway. Moreover, local managers had peacetime experience in running the railway locally because of the ethos of the colonial railway. The local stationmaster, Loco Foreman or Traffic Manager might have to play many different roles in support of a skeletal administration that keeps the railway system running on an economy. There was thus expectation, reinforced by years of personal experience, that the local railway manager would treat local and unpredictable exigencies with suddenness and decisiveness and without central direction.

6

The technical limits of the railway: The Mandalay to Lashio Branch

Details on the BR's layout of permanent way, its collection of engines and the location of its railway yards and loco sheds might be easily dismissed as superfluous technical data. On the contrary, the railways' technical aspect was as important to its performance in the short term as the topography of a landscape was to army commanders. Rivers and creeks, elevated areas and depressions, the existence or not of bridges, forests, hard ground or desert sands, or a range of other geographical factors limited and permitted what an army could do on the battlefield. Similarly, the technical aspect of the railway, its physical layout, equipment, resources and every other non-human element required to make the railway network and the trains on it work placed constraints and opened up capacities for the railways' ability to meet the demands placed upon it. The technical aspect could be adjusted (a river could be bridged or a hill could be levelled or tunnelled through) with an investment of time, manpower and other resources. But in response to the sudden onset of invasion conditions and demands that fluctuated in scale and positionally on the network, the railwaymen generally had to work with the existing technical aspect, however much this could be tweaked where time, materials and manpower allowed and all of these were in short measure in the FBC. These limitations came to bear heavily on the BR when attempts were made in March to develop the Mandalay area as a base area to replace Rangoon and then in April when yet another alternative had to be sought to replace the smouldering ruins of Mandalay.

Goddard had explained in late March that Rangoon was too far forward at the beginning of the campaign. A good base sufficiently behind the zone of operations would not need to be moved. He argued that what should have been done was that base installations should have been sited in Upper Burma before the war broke out. This view does not sit well with the available transport facilities in the north.[1] One legacy of colonial railway investment, for example, was that other than Rangoon there was little else suitable to serve as a major railway hub. Despite their reputation for, or perhaps merely the image of, grand investments in undeveloped terrains, colonial railways such as those in Burma were built quickly, with slim budgets and considerations as to availability of labour and other costings. Most importantly they were built for a purpose and while some lines might be for security, those in Burma were mainly built to serve the colonial economy, heavily built up, where demand required it, such as the reinforced Lashio Branch connected to the Burma Mining Railway at Namtu or the very long and fragile Myitkyina Branch. They were not built or designed for flexibility and they did not easily accommodate quick change in war. The railway's stations, yards and equipment followed the same pattern. Prendergast observed after the campaign was over that the railways were built to accommodate a cascading of traffic downward towards Rangoon and that facilities were scattered, of low capacity and in every other way meagre in the extremities of the network and increased in scale only as traffic demand increased as it grew closer to Rangoon. The FBC, however, was pushing everything from men, locomotives and equipment north, where it was difficult to find accommodation.

The present chapter continues the discussion from the previous chapter of the essential importance of using local experience and knowledge, in this case, the BR's technical and linguistic aspects, to inform Transportation Technique. The legacy of colonial railway policies had a real impact on their ability to conform to standardized procedures but also to support military operations per se. Among the more illuminating grumblings made by railway officers being shifted around the colony in April 1942 is the comment that no railwayman was consulted on the new location for the loco sheds, but that the new location was badly chosen. The problem was not a desperate attempt to complain about outside interference once again but a genuinely sincere plea to make use of local experience. The comment, made in a confidential report from the field to Biddulph, highlighted the continuing need to wed Biddulph's metropolitan military Transportation Technique with the local operating experience possessed by the BR men as discussed in Chapter 2. The colonial railwaymen knew the limitations of the BR's technical aspect, while others did not. Nevertheless, there were also things that local railway operators could use in operating trains in a wartime setting for which pre-war local railway operating had not prepared them.

Mandalay, Ywataung and Myitnge

After the fall of Rangoon, the complex of towns and stations around Burma's second-largest city, Mandalay, on the east bank of the Irrawaddy appeared for some time to make it a plausible alternative as an administrative centre for the railways. Mandalay was at one end of the Lashio Branch that was supposed to have been extended into China via the Yunnan-Burma Railway (YBR). A short distance up the railway line towards Lashio was Maymyo, the summer retreat of the colonial government, which had withdrawn here when it abandoned Rangoon. Myohaung junction to the south was now the most important junction in the colony, with railway headquarters (HQ) just across the Ava Bridge at Ywataung, and the second-most important railway repair facility at Myitnge. When Mandalay itself was devastated in the 3 April Japanese bombing raid, the railway complex around Mandalay underwent sufficiently extensive damage to warrant abandoning much of the Mandalay Station area itself. Attempts to keep the new railway hub in the north going by moving facilities to yet other locations were ill-informed and indicated that while rail operations could persist, the technical limitations of the railway network in the north, divorced from the network in the now occupied south, would be a major obstacle to these efforts.

The Japanese air attack on Mandalay on 3 April was both less and more damaging to the railways than the British or the Japanese military immediately realized. On the one hand, it was not as devastating to broader railway movements as one might have expected from the twisted metal of the railway yard, the destroyed communications controls and the burning embers of what had been the railway facilities other than the brick station buildings. The Japanese had misunderstood the railway arrangement in the Mandalay area. The major crossroads of the BR was at Myohaung and not at Mandalay; Mandalay had a large railway yard and many sidings for loading goods and was the HQ for operations. But in terms of the actual movement of trains on the system, Myohaung was far more important because it was the actual meeting point for four rail lines (the Lashio, Ye-U and Myitkyina Branches and the Rangoon–Mandalay Trunk Line). The Japanese attack on Mandalay thus did not disrupt railway operations as seriously as it would have if the Japanese had attacked the smaller Myohaung Station instead. After the destruction of Mandalay Station, the railway operations' physical hub was moved three miles south to Myohaung itself. As Myohaung had been the HQ for a district of the BR, it had the necessary administrative facilities, including the District Traffic Superintendent and Locomotive Superintendent's offices as well as those of other officers and staff. From this point, whatever traffic normally would have gone to Mandalay now had to be received, marshalled and dispatched from here. This made matters more problematic, for while it was an adequate administrative and communications hub, Myohaung lacked the railway yards of Mandalay, having only two full-length running loops. Hence, trains normally did not terminate here before 3 April and instead moved on to the railway yard at Mandalay.[2]

On the other hand, the 3 April attack had deprived the railways of a major support facility. While repair equipment was located at Myitnge, Mandalay had sported the facilities of a major railway centre. Now, damage to the railway yard was irreparable beyond the opening of two lines, one to the Mandalay riverhead and the other to the Fort Dufferin depots. Other than these two lines, the Mandalay Railway Station was abandoned as it was at a dead end, and left behind were a number of locos and some 200 railway wagons and carriages that had been damaged, burned or trapped in the railway yard by destroyed track.[3]

The turntable approach lines had also been damaged on 3 April, which meant that even though the turntable itself was undamaged, the turntable could not be used. Mandalay had no alternative means of turning engines and although a triangle had been proposed for Myohaung just two miles away, this does not appear to have materialized. Instead, the nearest triangle was at Ywataung, fifteen miles away from Mandalay Station.[4] It was thus decided to move the Railway HQ to Ywataung and to use it as the main locomotive depot. Railway personnel criticized the choice of Ywataung as the new HQ and loco shed for many reasons. First, there was no space. In the loco yard, there was only one line other than the triangle line. Further, half of the loco shed was being used at the time as a salt *godown* and thus there were only two lines available for locomotive use. Even the available lines were often blocked by wagons taking on salt. Second, fuelling and watering were difficult as there was only a single water column in the loco yard and no coal yard, which meant that in order to coal engines, they had to stand on the shed entrance lines, creating further blockage. Third, Ywataung had no stock of essential repair parts or stores and no place to store or issue oils. There were no tools and no facilities for washout. Ywataung also had no loco staff, only a pumpman, and no supervisory staff. There were no food supplies or quarters. For anyone who found a way to get to Ywataung, there was no provision to evacuate the staff from Mandalay. Fourth, it was difficult to fuel engines at Ywataung as it lacked any coal stages, baskets or shovels for the first few days and then no coal coolies afterwards. Fifth, engines were sent out from Ywataung but many never came back, in particular those that moved out on the Ye-U Branch.[5]

As on other sections of the line, the railway repair staff at Myitnge had to be held together through the leadership of Jack Cardew. Among their many complaints, voiced since early March, were the limited food supplies available, the lack of pay at a time when they needed to pay for their families' tickets out of the colony and a reluctance to keep working in wartime conditions that included bombing. Cardew improvised and initiated a number of reforms that were designed to ameliorate the situation. Cardew organized a Railway Provision Shop at which workshop and other railway staff could purchase their own food. He gave railway staff two months' pay in advance out of a special fund.[6] Nevertheless, staff went on strike on several occasions and some began to desert, fleeing Myitnge altogether. Ultimately, a platoon of Burma Rifles soldiers were ordered to Myitnge by an unsympathetic officer with orders that workshop staff were not to be allowed to leave Myitnge. The Burma Military Police (BMP) commandant of Mandalay gave orders to shoot the strikers if they tried to leave, and reportedly, Dorman-Smith gave orders for the Burma Rifles troops to shoot one in ten of the striking workers and to 'chuck 'em' into the Myitnge River afterwards.[7]

Morale among the railway workers at Myitnge continued to dip, making the mobilization and retention of labour for the railways and the railway shops a continual challenge. Many essential staff needed in the workshops jumped on trains going out at night. By 14 April, there was no staff at Myitnge to work the trains. Locomotive superintendent Cardew put together a virtual press-gang to go through every wagon in the railway yard to gather men each night.[8] Recouping staff confidence and hopefully their support required change above the level of local manager. A number of BR men believed that the military administration of the railways no longer worked and that the railways had to be removed from military control.[9] Hutton was also in favour of removing the railways from Biddulph's control, for although it was not a negative reflection on Biddulph's management, everything had to be tried to improve the transport situation in those parts of Burma that remained in the army's hands. Biddulph, who, as mentioned, had been *ab initio* aware of the challenges of the job, again asked Rowland if he would take over as Director of Transportation. Rowland refused as he felt it was too late to make any difference and he did not feel comfortable taking over the management of IWT included in the broad mandate of the Directorship of Transportation.[10]

Rowland eventually did bow to the inevitable, for the army needed to relieve itself of the burden of managing a railway that was mainly a concern for the colonial government and Chinese forces evacuating north and the latter two interests needed the aid of someone who could give the railways their undivided attention. Rowland finally accepted the case that his taking over would slow the desertion rate among railway staff.[11] However, he would only agree to step in after General Alexander agreed to a list of seven demands made by the subordinate railway staff. First, there must be an immediate evacuation of all of the railway families to India. Second, the railways must be demilitarized. Third, a vulnerability allowance must be paid to railway workers who remained. Fourth, the railway workers were to receive two months of pay in advance so that they could pay for the evacuation costs of their families. Fifth, the railway workers were to receive free rations on the army scale. Sixth, the railway workers were to receive reasonable notice (seventy-two hours) of the final evacuation and to receive military protection while they were being evacuated. Seventh, the railways were to receive protection from air raids.[12] Alexander agreed to all of these conditions, this being perhaps one reason why Rowland would write in a confidential letter in September that Alexander 'was the only decent General Burma had'.[13] One of Dorman-Smith's advisers, John Wise, has related that the governor's inner circle was surprised that Alexander agreed so readily to these conditions at this late stage in the FBC. Wise surmised that Goddard and other military authorities knew that the conditions could not be carried out. Certainly, circumstances would soon force Alexander to break his promise regarding the sixth condition.[14]

The new arrangement does not appear to have been clearly spelled out to all concerned and there remains confusion about what Rowland's appointment actually represented in terms of the transfer of control. Dorman-Smith formally appointed Rowland as the railway plenipotentiary or 'the administrative head of the railway'.[15] From the colonial government's perspective, the transfer gave Rowland the 'dictator powers we so badly needed', empowering him to 'g[i]ve orders to everyone except the GOC [Alexander] and HE [the Governor]'.[16] At the same time, a rumour made the rounds among the BR men that Rowland had refused to take over the railways and instead merely agreed to act as an adviser to Biddulph.[17]

The truth was somewhere in between. The 'head' of the railways was now a kind of chimera, being both Biddulph and Rowland. Indeed, Biddulph and Rowland appeared together in person on the same day at the stormy railway workshops at Myitnge to get the men to work.[18] It was a brilliant, if constitutionally unsustainable arrangement. To the military, the BR continued to function under Biddulph's orders and Rowland was taken on as a special adviser. To the GOB and most importantly to the BR men, Rowland was now in charge of the BR and, reportedly, Biddulph had agreed 'readily' to Rowland being put into unquestioned control of the railways.[19] To sustain this situation, a number of documentary fictions were created. While Rowland had admitted that by the following day he had not yet received written authority from Dorman-Smith, he went ahead and re-delegated his authority to Biddulph in writing, explaining that at Rowland's request, Biddulph was remaining in charge 'of all operational work, on and from 16th April', as if he were, instead, Rowland's assistant.[20] Nevertheless, surviving orders indicate that Biddulph would continue issuing instructions to Rowland on what Rowland should be directing the railways to do. Although it is sometimes assumed, even by some railway officers, that Rowland was pulled off the YBR Project to help manage the railways, he was not. During most of what remained of April, he both administered the Project and lent aid to the shrinking BR system at the same time, in his words 'dividing my time between the two, hare'ing backwards and forwards between Lashio and Maymyo'.[21] Rowland was not mandated to get the system running again on a permanent footing, but to help keep it alive only as a temporary measure. Evacuation from Burma was certainly going to happen, the only question was when. Rowland was only to lend his authority and reputation to keep railway operations going until the final withdrawal of the army from the colony. As he described what was handed over to him, 'It was just a question of holding on to the last possible moment with our army in full retreat.'[22]

From mid-April, rail operations would be dominated by evacuation operations. The civilian evacuation to the north of Mandalay was initiated after the Japanese bombings of Mandalay on 3 April and of Maymyo on 8 April. A few weeks later, soldiers, railwaymen and the civil administration were to receive orders to evacuate to the north of the Irrawaddy as well. With the last remaining aerodrome in Burma being that at Myitkyina on the other end of the railway line from Mandalay, the natural course for evacuees, both civilians and military wounded, was to move north in the hopes of flying out. At the same time, other military trains and some civilian evacuees were directed to the northwest to Monywa and Ye-U from whence they could walk out of Burma into Assam. Dorman-Smith estimated that the number of those evacuated by rail in the last phase numbered 15,000. Unlike during the Rangoon evacuation, however, few special evacuation trains were available. Most civilians were removed on military trains where they shared space with the families of several hundred soldiers every day. The military authorities allotted 250 seats on the daily evacuation train of five coaches from Myohaung to Myitkyina. On top of this, hundreds of others would take the train from Maymyo to Myohaung in the hopes of catching the train and thus getting out. All told, civilian evacuation authorities sent up to 500 to 600 people per day to Myitkyina alone.[23]

Nevertheless, trains moving out of Mandalay and Myohaung often well exceeded these formal caps on passenger numbers. There were opportunities to make small amounts of money from Indians desperate to get out. When the migration out of Rangoon began in late December, people leaving Rangoon were allowed to use the railway wagons on trains *gratis*, an odd measure given how much easier it would make the flight of the poorest residents and the men of the labour gangs who kept the port going so long as they remained. Stationmasters, however, often took bribes of varying amounts to let anyone on board who could pay the right price, but this hindered traffic flow and contributed to the congestion of stations and terminals not just of the railways but of the IFC jetties as well. At the latter, often hundreds and sometimes thousands of refugees encamped because they had made it thus far by train but there was no space for them to move onto the next mode of transport for the following leg of the journey. Things became so bad at Monywa Station that the District Commissioner of Sagaing banned all civilian travel on the railways on the Ye-U Branch, but, again, bribes meant that each train on that branch continued to carry 500 to 1,000 refugees.[24]

The railways were no longer kept up to pre-war standards or even those of the first few months of the campaign. To the BR men, rail operations now were almost unrecognizable. Rowland described the state of the railways as a 'frightful mess, everything … chaotic and beyond all hope of getting straight'.[25] Other accounts that attribute this state to railwaymen who had reached their breaking point confirm this appraisal. The morale of the railway staff was at low ebb. All aspects of running the railway had stopped working, no one would marshal trains, drivers could not be found or ran off when they were, supervisors no longer bothered to encourage or even to support subordinates.[26] It was in these challenging circumstances that Rowland was able to encourage the remaining railwaymen to roll up their sleeves, get back to work and keep the railways running for one more week.[27]

The GOB decision to relocate to Myitkyina and the increasing needs of the Chinese Army in Burma consumed available BR manpower and rolling stock. This meant that trains promised by the military for the civilian evacuation of Maymyo was actually afforded on only one occasion (although several additional, but unscheduled evacuation trains would be slapped together by railwaymen to get remaining railway workers and their families out on at the last minute). This train, the Maymyo Evacuation Special, carried 600 to 700 people, half of the train's ten coaches being allotted to the military. The train left the hill station on 18 April 1942 to deliver the evacuees intended for 'Air Convoy H'. Dorman-Smith would claim that in any case, even if there had been a daily train it still would have been insufficient to meet the demand to get civilians out and, moreover, air evacuation resources were unable to meet the demand for civilians who needed to be evacuated from Myitkyina Aerodrome anyway. Additional daily evacuation trains were promised after the first Maymyo Evacuation Special, but only if resources permitted, which it would seem they did not. As a result, as of 20 April, 4,000 people remained at Maymyo to be evacuated.[28]

Some Anglo-Burman refugees, who appear to have been forgotten by the civil administration and the military alike, wrote a memorandum to Dorman-Smith pointing to the loyal service they had given and also staying at their posts to the last possible moment, and now they were in danger of being left behind in the Mandalay evacuation camp. When Dorman-Smith's men investigated, a prominent Anglo-Burman from Rangoon pointed out that there was a large pool of Anglo-Burman engine drivers sitting in the camp who could take the trains, many of which sat unused, north. When the trains carrying the governmental departments had left Rangoon after the evacuation was ordered, there had been no reverse traffic of empties going south, so each Anglo-Burman driver who arrived at Mandalay at the beginning of March was simply released from service with the required one month's notice, entitling them to be paid for that period. These men now offered their services on the condition that they could take their wives and children with them, but were refused and told to take a boat instead. Nevertheless, the following day the governor acceded and the drivers were taken by bus to Myohaung Station where they put the trains into steam and took as many evacuees as they could carry on the journey north. Along the way, they would pick up Lady Dorman-Smith, the governor's wife, who was waiting at Shwebo Station also seeking a passage north.[29]

The Chinese withdraw

In the face of the tensions of the evacuation, it is likely that the behaviour of each side, British and Chinese, fed into the anxiety of the other. American and British accounts differ widely regarding railway problems at this time in the Maymyo hill section of the railway up to Lashio. From the American accounts, European railwaymen were unreasonably uncooperative and inefficient.[30] For their part, BR officials complained about the ongoing problems they were having with uncooperative Chinese officers. One of the main concerns of the Chinese Army in Burma as April edged to a close was to avoid being trapped by Japanese forces. Aside from the northwestern route up through Lashio, the only other route was the Ava Bridge across the Irrawaddy. Chinese officials, officers and soldiers likely also interpreted, as they had British withdrawals on the Rangoon–Mandalay Trunk Line about the same time, the departure of BR staff as cowardice or, worse, as signs of collusion with the Japanese. Chinese officers also objected to any more trains being sent west towards Mandalay because they wanted trains to go east to Lashio. As a result, BR staff at Maymyo were forced to keep rail movements and the departure times of evacuation trains secret, leading to further Chinese suspicions of British duplicity.[31] On 25 April, there was a fracas between the BR men and a party of Chinese officials and colonels over the despatch of a desperately needed Chinese troop train from Maymyo to Lashio. Milne called Rowland to the station to sort things out, the whole confrontation lasting from 2100 hours to 0200 hours that night. Milne records that 'it was then that these Chinese officials completely lost control of themselves and threatened to have us put up against the wall and shot'.[32] As Chinese engine crews were now available to take the place of BR men, the Lashio Branch was thus handed over to Chinese control on 27 April. For running the line, they were given three of the Beyer Garratt engines and nine Mallet engines that were necessary for the steep hill sections.[33]

Dorman-Smith argued that the behaviour of Chinese troops during this period made conditions on the line much worse and added to the confusion. They did so, he asserted, mainly by insisting that trains on the line should run according to their schedule, without consideration of normal or special timetables. They defied advice, 'gaily ignor[ed] signals' and moved without first picking up line clears. When railway officials objected they had 'a bayonet or revolver … poked into their ribs'.[34] While Chinese officers and soldiers, like many others, indeed resorted to doing desperate things in the race to get out, such evidence presented out of context unfairly suggests that all of the blame for the chaos of the Burma evacuation and retreat after the loss of Mandalay should be attributed to the Chinese forces. There are indeed examples of individuals and groups acting out of self-interest in attempting to survive in an increasingly desperate situation. As we know from the account of the Burmese Movement Control Officer (MCO) at Myohaung in April, many of the Chinese officers were seizing trains under orders from their superiors. They instructed their subordinates to seize any locomotives they found and anyone manning them as their only way out, and were doing so with the sympathy of the Chinese Liaison Officers (CLOs) who helped to intercede on their behalf. Lieutenant Colonel

Willis explained the situation in terms that seemed reasonable to MCO Law-yone. As the number of Chinese officers who tried to seize trains increased to the point that they began to interfere with railway operations, Willis and Law-yone worked out an arrangement in which Law-yone put together trains while Willis and the Chinese waited off the line, then Law-yone would call them up to take the train when it was ready.[35]

The Ava Bridge was the only railway and road link across the Irrawaddy River and thus promised to create a serious bottleneck in the movement of the imperial and Chinese forces and refugees between Burma north and south of the river (see Map 6.1). In anticipation of this becoming a huge problem in case of retreat, measures had been taken earlier in April to build ferries to conduct traffic across the Irrawaddy at that point. A ferry had also been established across the Myitnge River in case the bridge there too was lost. When 1st Burma Division crossed the Irrawaddy River, for example, it did so by ferry not by the Ava Bridge.[36] Even so, rail traffic had to move across the bridge. Dorman-Smith identified the period just

Map 6.1 Burma Railways South of the Irrawaddy, end of April 1942.

before the army's withdrawal over the Ava Bridge across the Irrawaddy River at Mandalay as the 'railway's greatest trial'. The main complicating factor was that with Mandalay Station out of commission, all remaining traffic had to move through Myohaung Junction, what Dorman-Smith called 'the Myohaung bottleneck'.[37]

Despite the claims of railwaymen at the time, the migration of both the military and the civil government across the Irrawaddy River on 26 April produced a traffic surge that worsened the functioning of the railway to a far greater degree than any problems produced by Chinese military users of the network. When Alexander decided that holding the eastern bank of the Irrawaddy was no longer possible, he directed the Advanced Army HQ to withdraw from Maymyo to Shwebo, north of the river. The entire area served by the Lashio Branch was also to be evacuated. This meant that Government House and the remaining civil government offices at Maymyo would also have to shift from Maymyo to Shwebo.[38] Rowland was thus given far less notice than the seventy-two hours he had been promised by Alexander to get 3,000 to 4,000 railway staff and their families out.[39] The Myitnge Workshops also had to be abandoned with Mogaung identified as the new site for a workshop. Ninety-eight wagons were loaded up with machinery, equipment and stores and sent up the Myitkyina Branch.[40]

The huge sudden surge in traffic demand produced many complications for remaining railway operations on the eastern bank. First, significant congestion and slowing of the network occurred at many points. On the same day, for example, there was a ten-hour jam (some trains waiting there for twenty-six hours) at Meza, which had twelve to fourteen trains 'confronting each other on the six lengths of line and sidings'.[41] Second, railway managers found it difficult to move through the network to new positions on the western bank while at the same time they were either stuck on the line or off it and could not help in relieving congestion on the line. Biddulph thus spent 26 April moving up and down the line between Mandalay and Maymyo assessing the situation and found the stations on the line lifeless, while senior BR managers including Rowland and Milne drove to Sagaing by car.[42]

Concerned with captured railway line and materials aiding the Japanese advance, BR men and others undertook efforts to worsen the limitations of the technical aspect of the sections of the railway that they were leaving behind. These included the same kind of denial operations that they had applied when they had abandoned the southeast, with little success, and Rangoon, with a little more success. But as the campaign had progressed, the defenders had become increasingly adept at leaving behind for the Japanese ruined bridges and twisted rails. Denial efforts regarding the workshops at Myitnge, for example, were more successful than they had been when attempted at Rangoon a month and a half earlier. Arthur Johnson who had directed the demolitions of the railway workshops at Rangoon seven weeks earlier was again called upon to take charge of the demolitions at the Myitnge workshops. The demolition teams needed three to four days to prepare for the destruction, the violence they directed at the machinery took twenty-six hours to complete. This included disabling twenty-four locomotives undergoing repairs by having their boilers burnt out, smashing their fittings and removing their connecting rods and sending these off to Kanbalu. A little over an hour after the orgy of destruction was over, the remaining locomotive staff at the Myitnge workshops were evacuated on a special on 29 April, reaching the other side of the Irrawaddy River the following day, but not Myitkyina until 3 May.[43]

Coward remained in control of the railway yards on the east bank of the river and directed the last of the trains that would cross the Ava Bridge for another decade. More immediately, this traffic represented the final waves of retreat, but they came from multiple directions and all at the same time. Dorman-Smith focused special attention on the evacuation trains that were still being brought down from Lashio, the Chinese armies withdrawing from the area of Yamethin and the traffic from the Myitnge Workshops, but this hardly captures the chaotic mass of humanity that was racing on the rails to cross the bridge before it was gone.[44] Traffic across the Ava Bridge in the last few days was unusually heavy and Coward only had a skeleton staff of officers and subordinates to marshal the trains and send rail traffic across it by this time. The crossing was a bottleneck not only because of the bridge but also because of the limitations of the junction. When the trains came in, they came in from the south with their engines on the north end, but the route across the Ava Bridge was to the south. Therefore, the engines had to be uncoupled, run around the station and then reattached to the south end of the train, so that the engines could pull them across the bridge. In a ten-hour period, from the evening of 27 April to the early morning hours on 28 April, thirteen heavy trains moved through Myohaung to the north before the Ava Bridge was blown.[45]

Chapter conclusion

The technical aspect of the railway can offer capacities to deliver the flexibility needed or place severe constraints on the ability of the railway to meet changing demands. In Burma, the railways were built to serve the colonial economy and little else and as a result, the railways in place when the Japanese invaded did not allow significant flexibility outside of the Rangoon area so that when Rangoon was lost, no single realistic alternative was available as a railway base area or hub. An effort was made to develop a broader assortment of railway facilities in the Mandalay area into a new railway hub. However, once the Japanese bombed Mandalay on 3 April, railway facilities became decentralized. They stretched from Ywataung on the west bank of the Irrawaddy to Myitnge to the south of Mandalay and to the northeast at the hill station at Maymyo. Finally, when this area was evacuated on 28 April, there were hopes to move operations to Mogaung but these were built on wishful thinking and little else.

Alongside the technical aspect was what might be called the railway service's linguistic aspect. Railwaymen in Burma were trained to speak the languages relevant only to colonial users and the heritage of connections with India, only severed politically in 1937, was a railway service that could only communicate in English, Burmese and a collection of languages of the Indian subcontinent and not at all with the armies of its giant neighbour, China. Unfortunately, Chinese officers and men who had become the railway's primary users on large parts of the network generally did not speak English and as discussed in the previous chapter were unfamiliar with a relationship between military authorities and civilian transport operators in which the latter made the decisions regarding when and to where railway engines moved. Help from the CLOs although frequently useful remained insufficient to avoid increasingly volatile situations in railway operations.

The problems regarding both the technical aspect and the railways' linguistic aspect influenced developing perspectives on the right way to conduct railway operations in a war zone even in the desperate days of late March and April. It was clear that however much training at Longmoor and handbooks derived therefrom could prepare military engineers to grapple with the challenges of running a railway under battlefield conditions, they could not make up for local knowledge about running a railway in a non-Western, colonial environment. At the same time, local knowledge had its limits and military training, including metropolitan manuals, warned operators that they might indeed have to operate trains where they could not communicate with users of the network. The need to consider both was among the first lessons taught at the new Transportation Training Centres in India in the months after the end of the FBC.

7

Dark territory: The Myitkyina Branch, late April to early May 1942

The withdrawal up the Myitkyina Branch brought together various parties of railwaymen, both colonial and military, who had been operating in different sections of the railway and who had their own local experiences with operating the railway during wartime conditions. This experience in operating autonomously of regular, central direction, as discussed in Chapter 6, was especially useful for grappling with the problematic conditions on the Myitkyina Branch. As this period progressed, the railway system became less discernible; the autonomous movement of single trains making use of the assets of the railway after the network had broken down defined railway transport. These problems were so frequent, dense and localized that small clusters of railwaymen could only keep parts of the line operating in small, autonomous clusters. The railwaymen had to do so because the Myitkyina Branch by this time was so denuded of station staff, its stations and thus its communication network so devastated by air attack and other problems, and alternatives for communication so absent (partly because of accidents and extreme congestion). This was dark territory, the term American railwaymen applied to sections of the railways in which signals and communications were absent.

This chapter examines how small groups of railwaymen maintained local railway operations throughout the last weeks of the FBC. The challenges and obstacles of this period were an omnibus of all the problems seen in the campaign for railway operations. The period and the continued performance of the railways on the Myitkyina Branch helped to assure that attention would be paid by New Delhi to the importance of the railways. This made it possible to keep the 'failures of the railway as catalyst for change' approach Biddulph and Manton had adopted since February as a means of keeping up pressure for the militarization of the railways in India. The period also reinforced why, as we shall see in the following chapter, Biddulph and Manton were so committed to recruiting the BR men for military training at Deolali and their mobilization for military railway operating and construction companies in India afterwards.

Collapse of order

The colonial government was led to believe that the army's intention was that the withdrawal from the east bank of the Irrawaddy would be to Myitkyina, leading civil evacuation to focus on that northernmost point of the BR. As mentioned, Alexander briefly entertained that thought, the army seeing one of the advantages was the possibility of keeping the Chinese divisions supplied with rice from the Shwebo area, necessary after the loss of the Kyaukse area to the Japanese. Soon after, Alexander changed his mind and withdrew his forces to the northwest towards Kalewa instead.[1]

By any realistic assessment, however, north-western Burma could not seriously be considered as a military base from which to resist the Japanese for any length of time. There had been no time to develop the infrastructure and without road or rail connections, keeping any forces of sufficient scale in supply over the course of the following months was implausible. What had never been expected in peacetime nor for a considerable period of time after war with Japan had broken out was that the railways would be pushed 'up against a wall' in the north. The Myitkyina Branch was a low-capacity line stretched across a vast territory, with railway resources scattered in tiny stations that had little potential to support the army. Nor could it handle the huge flow of up-bound traffic that remained of the BR: the collected locomotives, wagons, cranes, boxed railway equipment and stores saved from Insein and Myitnge, railwaymen, railway staff and their families, along with thousands of evacuees and wounded, government stores and records, military and hospital equipment and personal effects beyond description. This juggernaut now moved up the railway line to facilities that were designed to accommodate a downwards, not an upwards, cascade of traffic. At the end of April, the evacuation broke down into a desperate, all-out race to get out before the approaching rains hit the steep jungle slopes over which evacuees would have to walk and climb to get to India.[2] People of all stations in colonial life struggled with armed troops to get a space on any train north.

The railways would undertake to keep trains moving, but this could only be for a short period as there was no longer any way to keep railway equipment in good repair. The wagons sent up from Myitnge bearing the essential repair equipment were forgotten or otherwise lost in the confusion, and even if they had not been, it would have been extremely difficult to set up an effective repair facility at Mogaung, as planned, in such a short time. It is likely that like so many other plans in this time of confusion and panic, these plans had not been thought through very well.

In the far northwest, a local colonial official arranged for daily special evacuee trains to carry refugees at the refugee transit camp at Ye-U twenty miles from Ye-U Station to Kanbya Station on the line towards Monywa. The village headman at Kanbya in turn arranged for numerous bullock carts to pick the women, old men and small children among the refugees up at Kanbya Station and carry them, while the others walked, to the Chindwin River. Ultimately, thousands of Indian refugees were moved by this relay. However, it was a difficult effort, not made any easier by the refugees. On one occasion at Kanbya Station, refugees even held a sit-down strike as they felt that not enough bullock carts had been provided. There was also a continual fear that the Indian drivers would not stop at Kanbya but take their trains all the way to Monywa from whence they would abscond to get out of Burma themselves. Armed police were thus posted in the driver's van to make certain that orders were carried out. A Japanese move up the Lower Chindwin finally ended railway operations on the Ye-U Branch on the morning of 1 May.[3]

The railways had become a 'dying concern' as far as imperial military demands went. Biddulph could thus concentrate on *ad rem* transportation operations necessary for the army's retreat.[4] Biddulph's role as Director of Transportation, Burma, now effectively involved only road and river transport. The railway would again be left to local managers to fend for themselves on the basis of local conditions, but this was obstructed everywhere by the condition of the railway. The kind of personal initiative that would otherwise have been directed at confronting the worst challenges to operations thus far seen in the conflict was thus decreasingly realized. A good example of this was the peregrination of Chinese Liaison Officer (CLO) Frank Donnisson. After seeing his men off at Monywa as he had promised them, he returned to Shwebo on 29 April to help lend a hand to those not yet evacuated, Brigadier Martin explained to him the impending crisis on the Myitkyina Branch. Twenty thousand Chinese soldiers, Martin predicted, were going to try to board trains on the Myitkyina Branch to escape, but as there was no food anywhere along the lines, they would starve. The two men agreed to re-establish a working group of CLOs again on the Myitkyina Branch. Martin tried locating additional staff in Shwebo. Donnisson took a train to Myitkyina to recruit three or four tough men with detailed knowledge of the country who could help lead the Chinese armies out of the colony and to arrange for deposit of supplies of rice along whatever route they would take. Donnisson would spend the next four days trying to get to Myitkyina on a congested railway with 'the last remnants of imperial authority unravel[ing] around him'.[5] Martin for his part would take the CLOs he found and mustered at Shwebo and attempt to reach Myitkyina by train as well, abandoning the effort just north of Wuntho, at which point he and his party abandoned their train and headed on to Manipur on foot.[6]

Imperial forces and American sources would complain with dyspeptic disapproval about the conduct of the Chinese forces in general as they fled north to Myitkyina during this period, just as they had complained about their behaviour before the Ava Bridge was blown. How much a more robust CLO presence could have ameliorated conditions on the Myitkyina Branch in the week after the destruction of the Ava Bridge is doubtful, as the CLOs themselves had confirmed. Nevertheless, the breaking up of the CLOs certainly deprived remaining colonial authorities, including the BR men, of some of their only intermediaries with armed and frightened Chinese forces, to the detriment of effective railway operations in the first week of May. As one witness summarized the behaviour of the Chinese troops after the destruction of the Ava Bridge, they increasingly ignored proper railway operation; took control of trains by force; threw nurses, refugees and wounded British soldiers from trains; freely threatened to shoot senior railwaymen; placed such men under arrest; and were directly responsible for several deadly accidents.[7]

One poorly understood episode is that involving General Luo Zhuoying at Shwebo. At the end of April, the entire Myitkyina Branch had become a highly congested line with upward pressure growing from the far southern end. Fears of capture by the Japanese and being trapped behind large numbers of refugees and wounded alike made the seizure of locomotives and rolling stock an attractive option for at least some Chinese commanders. In the early morning hours of 1 May, Luo seized a train, including an engine and seventeen cars waiting to take the 2nd KOYLI's depot from Shwebo to Myitkyina, at gunpoint and then used it to move his men north that night without orders. Shortly afterwards, there was a train collision about twenty-five miles north of Shwebo and Luo's journey north was seemingly blocked. The coincidence of Luo's train being so close to the scene of the accident and anxiety and distrust among Stilwell's staff, compounded by the confusion and poor intelligence available at the time, led to the canard that it had been Luo who had caused the crash. The headquarters (HQ), American Army Forces and HQ, China-Burma-India Theatre of Operations had not been happy with Luo's actions and when they heard of the crash they believed Luo had blindly crashed his train. The same version of the story crept into different documents during that period and as a result, all accounts published on this period that mention the accident blame Luo.[8] None of these accounts appears to have had access to the personal and very detailed account of Col. C. G. Stewart, the Chief of the Rangoon Police, who was on Luo's train at the time and who personally examined the aftermath of the crash. The crash had resulted because the decline in rail administration no longer afforded reliable signalling. After halting because of an air raid warning, the Chinese railwayman, Tseng, had sent a smaller train ahead of Luo's north, possibly to avoid both trains being caught in the same place by Japanese planes. The small train and another downwards train, the latter perhaps bringing empties south to fetch refugees, were both travelling tender-first without headlights and this resulted in a 'tender-on' collision that wrecked the locomotives of both trains.[9] A gang of Chinese railway veterans among Luo's men were then put to work to force the locomotives off of the tracks, make the necessary repairs to the line and clear the line to allow the movement of traffic to commence.[10]

In the midst of the deteriorating situation, brave efforts to cooperate did continue, even successfully. As one officer reported to Biddulph on 29 April, Brigadier-General Sung, Mr Tang and Capt. Paul Jones were trying to keep the traffic moving between Ywataung and Myohaung up to the Kanbalu. They did this by improvising arrangements, the four of them personally travelling on the engines, making certain that at least one of them was always present at Shwebo. However, by this time, all the engines on the Myitkyina Line north of the Ava Bridge were north of Shwebo and no one had brought any trains back down since the previous day. When Brewitt returned to Kanbalu on the morning of 29 April, he promised the group he would send down more locomotives to keep the traffic going and later in the day Jones and Sung took a jeep north to try to speed things up. Worse, no BR staff remained at Shwebo. A Signals officer's attempts to contact Kanbalu by telegraph elicited no response. In short, whatever was happening on the southern section of the line was unclear even before the abandonment of the east bank of the river.[11]

At the time the Ava Bridge was blown, Biddulph sketched out a breakdown of the railwaymen who needed to be withdrawn and estimated that there were thirty at HQ, 400 in the militarized railway companies and about 1,000 civilian staff all of whom would have to be withdrawn either via Myitkyina or Ye-U and Kalewa.[12] Although this staff would have been sufficient to handle traffic on the line in earlier weeks, the widespread congestion on the line, the lack of signals and repeated railway crashes and derailments, as well as Japanese air attack made this work difficult. Worse was the collapse of civil and military order on the line. As both the colonial state disintegrated and the Chinese retreat had taken on the form of a rout, the major users of the railway no longer followed rules of any kind and revealed vividly the problems of civilian railwaymen operating in Front Areas without weapons, despite being officially militarized.

The Chinese, American and remaining colonial railwaymen from the south who pushed north across the Irrawaddy onto the Myitkyina Line were now exerting authority over sections of the line. Confrontation and problems nevertheless lay ahead as these two groups of railway users and operators collided. A good example of what the confusion could do is that of the ambulance train that the BR had lent to the Chinese. On 28 April, J. S. Vorley encountered this ambulance train and found that it had been stuck at Tantabin Station for ten days.[13] On his own authority, Vorley had the Indian stationmaster telegraph for an engine to be sent down the line from Kanbalu. This engine would arrive the next day and the hospital train was now taken north back to Kanbalu. When they reached Kanbalu, Vorley's car was shunted onto a different train just ahead which was there to take the railway staff north. When the Chinese patients saw this, those who could still bear the pain grabbed their rifles and hand grenades and surrounded the engine of the departing train, threatening to 'blow it off the rails' if the train budged. Vorley had no choice but to allow the Chinese train to leave the station first, the train with Vorley and the evacuating railway staff leaving afterwards. As Vorley admits, the Chinese soldiers thought they had little choice but to undertake this action, for with few hospital staff on board other than Dr. James Wallace Lusk, continued delays in medical treatment meant death for a large number of them.[14]

The problems with the Chinese Army were no worse an obstacle than the range of other problems the BR men had had to deal with in operating the railways in a campaign that had gone progressively worse for them with each passing week. However, the problems with the Chinese Army could have been much worse had both sides not relented and accepted the collapsing scene around them. What had been demonstrated, however, was that if the alliance between countries with such vastly different cultures of operations between military and civilian authorities was going to do any better in resisting the Japanese and eventually push them out of Burma, much more work had to go into developing mutual understanding and cooperation. This work would be undertaken in India over the course of the next few years.

The end of the line

Biddulph saw the line congestion on the Myitkyina Branch during the first week of May as the last blow to railway operations in the colony. He described it in his report after the campaign as having brought all traffic to a 'standstill'. Goods and coaches blocked all sidings, trains overloaded with refugees packed the stations. Engines stood dead, unable to obtain water. Collisions and derailings blocked everything else.[15]

Figure 7.1 lends support to Prendergast's aforementioned assertion that the northern sections of the lines in Burma, built to accommodate the colonial economy, simply lacked the capacity to handle the rail traffic as it went north en masse as the campaign progressed. The space in the sidings, which is generally, but not always, about 250 yards in length, is not as important as their number, which indicates the capacity of the line to take trains off of the mainline and allow through traffic. Of course, many factors were involved, one being that when timetables broke down and line congestion ensued, the availability of a single siding in a section did not provide the same flexibility for traffic as it did when trains were coming through at equal intervals. Nevertheless, the decline in sidings available theoretically, leaving aside sidings already blocked by dead trains, provides an indication of just how far the flexibility of the line shrank

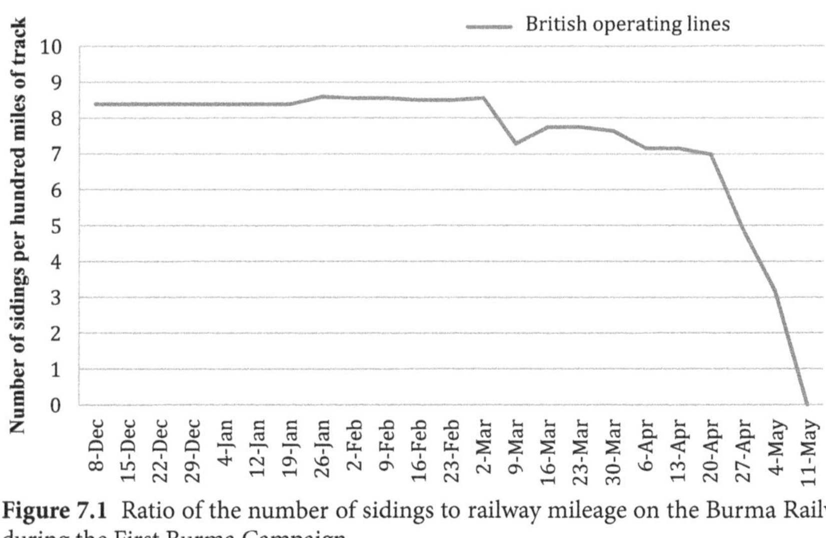

Figure 7.1 Ratio of the number of sidings to railway mileage on the Burma Railways during the First Burma Campaign.
Source: Based on figures derived from Biddulph, 'Report on the BR in Special Reference to Hostilities in Burma', 18–20.

generally. As a result, the potential for line congestion rose over the course of the campaign as locomotive density increased, as shown in Figure 7.2. Additionally, in the last week of the campaign, of course, most of the remaining sidings were filled. The mainline was in fact severely congested in many sections.

Both military officers and civil officials were having difficulty coping with the situation they were now in north of the river. On 3 May, Army HQ abandoned Shwebo and joined the retreat to Ye-U.[16] For the civil administration, it looked like there was little else to do, best captured by Dorman-Smith's almost lachrymal comment in his unpublished memoirs that at that moment, '[i]t was a sad Governor who sat in his house at Myitkyina without the remotest idea as to what fate had in store for him in the immediate future'.[17]

Telegraphic interruption made things much more complicated for railway traffic on the Myitkyina Branch after the neutralization of the Ava Bridge. The station telegraphic workers on the line had fled and, in some cases, the telegraph lines had come down between the stations anyway, leaving the remainder of the BR without telegraphic communication. Some stations had no staff left at all. Hence, the Absolute Block System had to be abandoned for the remainder of the FBC. On 4 May, trains were often operating without 'line clear' tokens on what remained of the BR, which by this time consisted only of the northern half of the Myitkyina Branch, although it does appear that several stations still attempted to negotiate these with the limited communications facilities available.[18]

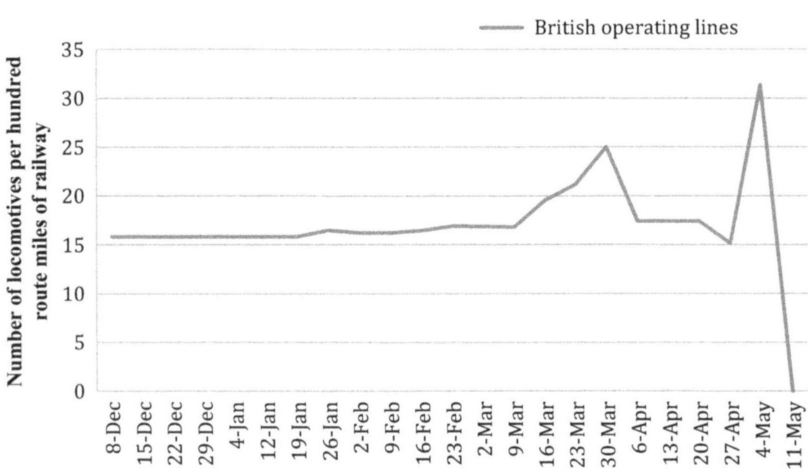

Figure 7.2 Locomotive density during the First Burma Campaign.
Source: Based on figures derived from Biddulph, 'Report on the BR in Special Reference to Hostilities in Burma', 18–20.

All of the problems related to overtly military–civilian conflicts in running the railways would manifest themselves again in many places and times during the railways' last few days, especially as obstacles and demands increased exponentially and the constant risk of death or being captured by the Japanese put everyone's nerves on edge. However, this period would also see the overlap on a shrinking stretch of line of many other problems discussed in the previous chapters. When all these problems intersected on the Myitkyina Branch in the first week of May, the rainy season was mercilessly and tantalizingly still a week or so away from its usual mid-May arrival, which would normally bring rain and a drop in temperatures. Instead, the 'railway corridor' became a hellish cauldron where stranded passengers baked in the sun, with as little water for themselves as there was for the engines. Many of the railwaymen still working the line had held out until very nearly the end, but now joined others walking westwards towards India. Rowland, Prendergast, Brewitt and other BR men commenced the walk overland with thousands of others on 4 and 5 May.[19] Of the roughly 22,000 railway workers at the start of the campaign, 11,000 attempted to walk out, and of these over 1,600 were eventually reported to have died. As Dorman-Smith probably very accurately observed in 1943, in the case of the latter, exhaustion from their extreme overwork on the railway 'for weeks on end' contributed to their deaths.[20]

Despite the end of railway operations, other transport agencies played a continuing but diminishing role for days, under Biddulph's management, until the evacuation became entirely a struggle on foot across the muddy hills in the midst of heavy monsoon rains and a maddening onslaught of malarial mosquitoes. Biddulph's personal notepad which he carried during the campaign includes a handwritten order from Alexander, dated 6 May 1942, explaining that Biddulph was to proceed to India (Assam) 'and make every endeavour' to help civilian staff (particularly railways and IWT) escape from Burma. Biddulph was to be given the highest priority regarding 'all forms of travel by land and air' to carry out these orders. On the same page, immediately below this order, and written in another hand, is another order, dated 7 May but with an unfortunately illegible signature, indicating that the execution of Alexander's order had to be delayed because of the seriousness of the situation in Kalewa, which presumably required Biddulph's aid to sort out. Several days later, Biddulph was still in the area and he received orders on 9 May that he was now placed under the orders of the Commander of the BurCorps and was tasked with ferrying BurCorps across the Chindwin. He did so from Shwegyin, along the imperial military force's line of retreat, from 9 to 11 May, using what remained of the IFC river fleet to ferry the troops that would move on and then scuttling the steamers.[21]

Chapter conclusion

The BR men were able to carry on local railway operations on the Myitkyina Branch until the end of the first week of May. This was despite the compounding of the accretion of all the problems experienced thus far in Burma during the campaign on a single stretch of railway that had been among the network's most fragile, strung out and least blessed in terms of railway support facilities. As the railway began to slowly stop, it was only the ability of local railwaymen to operate autonomously of higher levels of command that made it possible to keep even these limited movements going. It was also the ability of colonial railwaymen to bond together and their 'local' knowledge of the railway that allowed them to persist. This continued importance of local knowledge demonstrated that however much Biddulph's Transportation Technique was needed in India, it was just as important to wed it to the experience of these men who brought together the skills of operating a colonial railway in a non-Western environment made clearer even more than ever before by their perseverance independently of central control.

8

After Burma: The militarization of railways in India

In May and June 1942, weak and weary colonial railwaymen, many in uniform and with various ranks they had not held a few months earlier, crossed through malarial jungle and mountains and along muddy slopes as the rainy season hit hard. When they finally trickled into Assam, some found themselves in evacuee hospitals dying of cerebral malaria. These were of course the veterans of a valiant effort to keep railway operations going despite the retreat of the army and the collapse of the civil administration around them. In an ongoing war, however, they were most valuable because they possessed valuable experience in testing out Transportation Technique in a war zone.

How these men would influence railway operations in India after their evacuation depended upon a different kind of struggle waged by a particular group of individuals. Biddulph, Manton, Prendergast and others were able or perhaps were naturally oriented towards turning each experience into a test of what they knew and a possible lesson to mobilize in the training of others. They adapted what they knew to new circumstances and developed responses they then tested out or imagined ways in which they could be tested and based on observation and forward thinking were in a sense operational area innovators. When they believed they were right, they became reformers, actively advocating adaptation and change in particular according to the Transportation Technique they had developed and taught at Longmoor and had adapted to the experiences and conditions of not only Iraq and India but, above all, the ordeal in Burma. Before going to the Middle East and then to India, Biddulph had first passed through West Africa and had observed transport failings there, where Transportation and Movements staff were arguing for a larger establishment and more control, but their Command Headquarters (HQ) paid no attention. He had not said anything, because as he was merely passing through and it was not his country (or his assignment), he was a 'prophet with honour', and kept his opinions to himself until local staff asked him to do so. When he reached India and Burma, however, which he saw as his charge and thus his own country, he became a prophet 'without honour' and could lunge straightforward advocating change.[1]

Biddulph was indeed possessed of a profound belief in the importance of military Transportation Technique (and Movement Control, although not examined here). He preached an evolving and adaptive set of principles, procedures and approaches that changed with each new context and experience as it moved around the empire, from Britain to the Gold Coast, to India and then Iraq, and then to Burma and back to India once again. In each new venue, this approach found military and civilian resistance and colonial and military dogma. This chapter will examine Biddulph's and others' success in transforming the defeat in Burma into a catalytic event and in mobilizing human resources from this experience and channelling them into the service of Indian military transport. This ensured the acceptance of Transportation Technique and the need to militarize the railways in India. The many measures adopted by the railways on the Assam LofC were also in part due to the experience on the BR. These measures included not only decentralizing authority in railway operations and other administrative changes but mechanical, fuel and communications improvement as well. The acceptance of military Transportation Technique and the willingness to change in other ways how things were done, an acceptance made possible only by the failure of the railways in Burma, also helped to pave the way for additional changes, especially in Assam, that came with a greater American role in logistical support in northeastern India. The British would again have a colonial railway, with the operators and users of different nationalities, keeping an army supplied in war. This time it worked.

Blame and lessons learned

Broad consensus that the defence of Burma was not a success confirmed that Burma would not be turned into an Asian version of Dunkirk. It was squarely a defeat, as Alexander, Commander in Chief of the defence of Burma from March 1942, admitted in his memoirs when he observed straightforwardly that it was 'a complete military defeat'.[2] Indeed, as another commander involved in the campaign, Stilwell, famously remarked on getting out of India, his famous asperity reflected in his diction, 'I claim we got a hell of a beating. We got run out of Burma and it is humiliating as hell.'[3]

India underwent a range of reforms after the defeat in 1942 that helped to ensure the success of Allied arms in 1944–5. Among the most important were logistical changes that drew upon the lessons of the FBC. Dunlop identifies the key elements of large-scale logistical changes that ensued in response to three main problems or challenges all aggravated by limited resources. These included the limitations of India as a strategic base of operations, the limited capacity of the operational LofC and the inability to keep forces in supply that had for tactical reasons gone into the jungle and left the LofC. The resolution to these problems included building up India over the course of 1942–4 into a massive producer of military stores and supplier of manpower, on the one hand, and developing roads as well as the development of capacity to keep units in the field with supply using animal, water and air as means of transport.[4]

Railway operations also underwent reforms that have earned less praise or attention but owed their acceptance in India to the FBC. The difficult effort to introduce Transportation Technique to India became easier after the failures of civil railway administration in Burma. Biddulph kept the cause alive through letter writing and telegrams to key commanders in India and Manton worked the halls of GHQ India as well. Maj. Gen. W.E.V. Abraham explains that when he returned to India after the loss of Rangoon, Manton and Gardiner each approached him. They both emphasized to him in conversation that there was now an urgent need to militarize the railways in India. Manton's and Gardiner's points were buoyed by the fact that some of the same problems seen in Burma were being experienced in the Middle East and he was highly interested in a resolution to the problems of relying on civilian staff in war areas.[5]

Abraham now helped champion the cause of militarizing civilian railway staff in eastern India, specifically in Assam, which ran along the northern frontiers of Burma. Abraham subsequently (13 March 1942) wrote a 'Most Secret' memorandum, 'Railway Operations in Eastern India'. Abraham claimed it was 'foolishness' to believe that civilian railway staff would be able to perform in an invasion that has happened or one that may take place, observing that '[t]hey simply melt away'. Now that a possible invasion was coming to India's north-eastern gates, Indians and probably many Anglo-Indians as well would run away at the first sign of an attack and the railway would fail here too, unless special measures were undertaken. He suggested that the best solution would be for military ROCs to not take over the railways or sections of them because they could not run them as efficiently, at least not for some time, as existing staff, and there was an existing shortage of troops that made it difficult to supply such units. Abraham had clearly observed the need for local knowledge, but he also stressed the importance of military Transportation Technique and militarization of the railway staff per se. As he suggested, the solution was for the militarization of the existing staff on the 'threatened sections' and to subject them to military discipline. Only then could India depend upon such staff at their posts and it would mean the least dislocation to existing staff. Although there would certainly be a number of 'serious political and other objections', they had to be considered secondary to the possibility of a Japanese invasion, for if it took place without militarization, the railways 'would not stand the test'.[6]

The trend of opinion towards the militarization of the railways in India, at least those in operational areas, would not be reversed when Biddulph took on board Rowland from the middle of April. Certainly, after Burma fell, there was no longer any question that militarization was necessary for railway personnel in operational areas. As Biddulph, now in India, observed in August, 'the phase of "Is Militarization necessary?" is over'. Instead, the questions so far as forward thinking was concerned were how much time was left before the Japanese invaded, what difficulties could delay preparations and how could they be resolved?[7] These questions were being dealt with by transport personnel in India, including many from Burma as we shall see further below. However, this process occurred simultaneously with one of backward thinking as those military and civil authorities in command or in charge in Burma during the FBC began to piece together facts and opinions and make a case for who was to blame for the disaster. As militarization of transport troops was now accepted as necessary, blame depended upon identifying who was responsible for delays in implementing militarization of transport personnel and who mishandled implementation when it was put into action.

After having lost the campaign, history writing was also put to the task of narrating the defeat to help prepare the groundwork for the counteroffensive that would eventually retake Burma from the Japanese. The historian chosen for this task in October 1942, barely six months after the evacuation, was E. C. V. Foucar, one of those who had escaped from Burma. Foucar was commissioned by the army to write a history of the campaign within a six-month period, although the magnitude of this task required an extension of several more months, resulting in his draft of the *First Burma Campaign*. Foucar would publish a much more circumspect and less detailed account of the campaign, providing little more than chronological narrative of the events, in two sequential issues of the *Journal of the United Service Institution of India* in 1943.[8] Foucar's lengthier account of the campaign was never published, but is available in manuscript form in the British Library, the Imperial War Museum and in the School of Oriental and African Studies, the University of London's special collections, the last of which the present author consulted.[9] Foucar's mandate was to quickly assemble all the facts and put them together into an account of the failure of Burma's defence. The material Foucar used for his research included war diaries and other daily records and communiqués not lost in the campaign. The availability of this material was extremely uneven, especially at the time. Many documents had not made it out or were lost after being sent to India. A great deal related to the earlier parts of the campaign, especially from GOB departments, was destroyed to deny it to the Japanese, as they could not remove all of the material. Much of what remained burned up in railway wagons struck by Japanese bombs. Oral narratives and interviews thus played a larger role in reconstructing events than they might otherwise have played.

Unfortunately, some of the surviving correspondence between Foucar and participants in the campaign who reached India (he was writing in 1942 and 1943, so those who did not make it out and were captured, not to mention the many who had died, could not be interviewed for obvious reasons) that he took unequivocally were sources who apparently had strong personal, institutional and, in the context of the colonial period, racial biases. The railway's failures, for example, are thus frequently directed at the Chinese who joined Burma's defence. In some cases, it is argued that the administrative choice of personal rivals for wartime control of the railways was a contributing factor. It is unclear how many at the time had access to Foucar's narrative, but it was certainly used by Dorman-Smith who wrote in exile what would be the most influential account of the civil side of the campaign in 1943.

Ironically, some of the opinions directed by Burma colonial officials in exile, however much directed at Biddulph's management and away from themselves, nevertheless worked to support his agenda in India of keeping the militarization of transport moving at a fast pace. From the colonial perspective, for example, the late imposition of militarization and the limited nature of militarization were responsible for the declining performance of the railways. Too late and only a half-measure, militarization did not fare as well, perhaps, as the existing colonial management might have, had it been left in place. Foucar's and Dorman-Smith's accounts reveal a continuing tension between the 'military' (the non-local, imperial) way of doing things and the way things had been done under civilian administration. As we have seen, the constantly changing and simultaneous civilian and military demands placed on the railway required a precarious balancing act. Both Foucar and Dorman-Smith strongly believed that militarization was introduced too late to allow room for civilian railwaymen to adapt.[10]

The tone was also very much directed against the ways in which the defence of the empire had thus far been handled. Dorman-Smith's and Foucar's criticisms, for example, were double-barrelled and dug deep into the way imperial (rather than colonial) defence had been handled. As they constructed the causes for the railway's problems, a lack of *imperial* military resources had forced the *colonial* Burma Army to rely on *colonial* civilian railwaymen who remained un-militarized until it was too late. The empire, and India in particular, had few trained *imperial* military personnel to lend Burma to serve as a core of proper railway operating battalions. Dorman-Smith in his 1943 secret despatch on the fall of Burma, meant for official eyes, explained that their surrender of control of the railways represented the military 'admitt[ing] their helplessness to control the situation'.[11] Without the evidence provided by Biddulph's privately held documents, the near silence from the imperial military records on the reasons behind the transfer of control from Biddulph to Rowland would have otherwise seemed to confirm, incorrectly, the colonial view.

The colonial case against Biddulph in particular was argued vigorously. The fact that the army would write no official history of its own of the events of the FBC until long after the war meant that at the time, in 1942, it was easier to push Manton and Biddulph out of the picture than it was to arbitrate who was actually to blame for the failures in Burma. Perversely, the man who was most responsible for keeping the railways going in Burma during the campaign and who, along with Manton, would be so influential in the evolution of modern military Transportation in India was blamed for his difficulties in working with civilians, however much all parties respected his administrative skills regarding military Transportation. Biddulph would eventually be shunted off to Sri Lanka (at the time Ceylon) where he was made Director of Transportation over that colony which, as the HQ of the South East Asia Command, indicated that his departure represented no exile from the centre of activities. Manton was also to leave India very soon after the end of the Burma evacuation. Brig. R. Gardiner filled his vacated position at the Transportation Directorate.[12]

By contrast, Indian Army accounts also tried, sometimes unfairly, to blame the colonial government's performance for the railway's failure. Christopher Bayly and Tim Harper, for example, point to the testimony of colonial officer Donnisson who learned from the army's public relations officer, drunk at the time, that he had been told to turn Dorman-Smith into a scapegoat and blame him for everything that went wrong in the campaign. The purpose of this was to keep attention away from the army's own failures.[13] The US Army was also a party to the evacuation and to railway operations in it, as we have seen, and similarly its histories of the FBC, which closely follow the accounts and data of American participants including Stilwell himself, blame the colonial government directly or indirectly regarding the failures of the railway. A US Army account of the FBC, written for internal use in 1944 and distributed among Indian Army analysts, explicitly directed the lion's share of the blame at the Dorman-Smith government. As it argued,

> The complete refusal of the Civil Government of Burma to accept the juggernaut that was rolling up to cause its destruction probably represented the worst form of bureaucratic colonial government to be found. The general attitude of the officials as exemplified in poor roads, inadequate railways, private monopolies, the failure to act in all crises, can best be summed up by stating that they simply could not be bothered. Later Dorman-Smith displayed well-mannered concern, but he refused to declare martial law because 'it would disturb the native population.' Apparently bombed and destroyed towns and a general flight of the people were not considered disturbing.[14]

In their history of Stillwell's mission to China, as well, Romanus and Sunderland use the word 'maladministration' only twice. On the first occasion it is in reference to the Burma Road, but on the second it refers to those aspects of colonial Burma that made the easy flow of traffic through Burma difficult, from customs and transit formalities, to time-wasting regulations and major bottlenecks in the railway network.[15] The authors' observations about the performance of the railway during the FBC was simultaneously succinct, comprehensive and damning, placing the blame on the GOB and the railway management for their resistance to wartime measures and lazily maintaining their pre-war routines.[16] This view reflects the tone of Stilwell and his staff's opinion about the BR men. While this perspective is not unfounded, it does not fairly render the circumstances in which the railwaymen found themselves in 1942, nor does it capture the diverse reactions of the railwaymen to the call to duty during the campaign. However, it does suggest three things that are demonstrable based on a wide range of testimony and archival documents. First, the colonial railway management, like the colonial administration, was not eager to involve itself in a conflict outside of Burma or to allow outsiders to involve themselves in railway operations inside the colony. Second, there was dissension between imperial military authorities, and colonial authorities that would cause considerable delays in responding adequately to the Japanese invasion. Third, the railwaymen in Burma were unprepared for modern war, which arrived first with Japanese bombers.

The idea of a defeat preliminary to a future campaign had clear utility during the war. It was the case that acceptance of that campaign was a convenient intellectual device for mobilizing those in India to undertake changes out of fear. Once everyone accepted that the campaign was a defeat, whoever was responsible, recovery and improvement and preparations for ousting the Japanese from Burma later on in the war could be undertaken. Until the Japanese attacks of 7/8 December 1941, colonial military hubris in Burma among many other factors contributed to the failure to take Japanese military capabilities with sufficient seriousness to prepare adequately. Before the Japanese could be pushed back out of Burma, an admission of utter failure had to be accepted and explained. The defeat paradigm certainly provided a good basis for the caution with which the Allies over-prepared for the campaign of 1944–5, a caution reinforced by repeated failures in Arakan in the intervening years. The defeat paradigm was also a necessary complement to the victory narrative of 1944–5, captured in the title of Field-Marshal William Slim's popular account, *From Defeat into Victory* (1956).[17] Certainly, by comparison, 1944 to 1945 was better for nearly every Allied element involved other than the railways, which was an entirely different network in Burma by that time, held together by makeshift repairs and powered by jeeps rather than locomotives.

The FBC became or, rather, was transformed into a catalytic event that had the potential to cause military and civil authorities alike to take heed and accept that a commitment to the militarization of all civil transport agencies in India was necessary, however incompletely this would be realized. However, it required translation of the defeat into a lesson before it could be channelled into adaptation because of the politics within the Indian military. A key factor that made this translation a success was command change in India. The new QMG, Vickers, was already on board with the call for change from the start. As he observed on 25 August 1942, a lack of training and experience had caused India to be late in 'implementing measures which had been advocated … already cost[ing] us many thousands of lives and prisoners and much unnecessary hardship both to troops and civilians'.[18] However, the same kind of foot-dragging that had slowed the pace of militarization and mobilization in Burma eight months earlier still threatened to undermine the defence of India. Although in May, the QMG had called for all three of the transport services to be militarized, for example, four more months went by without change except for the railways. At that point, the Japanese bombed Calcutta and stopped work in the port, recalling Rangoon, now in enemy hands, in late December 1941, when all work stopped there too. This should have reinforced the message to everyone that Burma was not an anomaly; it was a premonition of things to come to India if changes were not made quickly to all aspects of transport.[19] Still, foot-dragging, however less effective it became as an obstacle to adaptation and change, continued, slowing the pace of the evolution of military Transportation in India.

Before he left India for Sri Lanka, Biddulph spent many months in different capacities relative to Transportation and Movements. By November 1942, he was employed as Special Inspector in the Department of War Transport because Transportation Technique remained inadequate in India and 'much waste is taking place'. This limited role involved mainly investigating how the Indian Army was utilizing transport and mainly involved making reports on wasteful usage by the Army Services in the War Department to other departments. As Biddulph lamented at the time, this 'naturally means a good deal of opposition'.[20]

Biddulph remained convinced that the 'Indian transportation efforts' were of 'tremendous potential importance'.[21] Biddulph thus continued his efforts to encourage metropolitan military influence in India, perhaps to prevent the much feared backsliding that would undo gains then in the process of being made. In November 1942, he wrote to Maj. Gen. D. J. McMullen, the Director of Transportation in the War Office (WO) in London, urging him to come out to India and take charge of Transportation and Movements or at least visit and presumably lend his authority to continuing reforms.[22] Biddulph's efforts also included an assessment of whether or not the transport effort in the FBC had been a success or a failure. On the whole, he argued Transportation had more equipment than was necessary for the job, as evidenced by the wagons that had been jettisoned. Transportation served the military well at all the critical moments, such as during the withdrawal from Rangoon. There were also sufficient civilian transport personnel, when bolstered by 'improvisations' from military staff, to do the job, and the defenders of Burma 'had far greater transportation capacity than the enemy throughout the campaign'.[23] The minor failures, including the absence of locomotives on the Kyaukpadaung Branch, Biddulph argued, did not influence the course of the campaign and even had Transportation been '100% efficient, the Army would, nevertheless, have been driven out of Burma'.[24]

Thus, in the chaos and retreats there were some things that worked, or at least worked for a while, despite serious setbacks and obstacles. The railway was one of these. From the perspective of December 1941, after the Japanese bombings of Rangoon, when much of the Indian railway labour disappeared and terminal congestion was heavy, or from that of February 1942 when the CRC claimed that the railways in Burma were effectively dead, the idea that the BR's story was indeed over would have been a venial assumption. Likewise, it again looked like everything was over for the railways when the Mandalay railway yard was destroyed by burning Royal Air Force ammunition railway wagons, followed by a firestorm that burned the rest of the city just three weeks after the evacuation and self-destruction of Rangoon, formerly Burma's main railway hub. These dramatic situations were joined by many other negative interventions that might have otherwise dealt a deathblow to rail transport operations, but the railways were atavistic. When everything else failed, and, again, the railway failed repeatedly, it revived and kept going, beleaguered and wounded, but still working, until nothing else was left.

Nevertheless, Biddulph suggested that the most potent factor in Transportation's successes and ability to keep going was the army's failures. It was only by the continuous withdrawal of the troops that forward stations were able to close just as they were about ready to crack under the pressure of bombing and desertion. If the army had somehow stabilized the situation and maintained a fixed Front Area, transport would have become insufficient and then withdrawal would have become inevitable. Even worse, in the case of a successful counteroffensive against the Japanese, it would have been very unlikely the railways could have supported it because not only would this have involved the operation of facilities, but it would have also required reconstruction, something the railways would not have had resources to undertake in the campaign. Biddulph was clearly making his case for strengthening the Transportation establishment in India. He added that, as terrible as the impact of Japanese bombing was in Burma, it promised to be worse in the case of India, as evidenced by the flight of railway staff at Manipur Road and on the Manipur–Lumding section. A higher percentage of Indian railway staff fled than in Burma, where the staff more often 'stayed put'.[25]

Biddulph's specific recommendations were outlined in a report he made on Transportation and Movements in India, but these represented the culmination of points he had been making with increasing clarity from the beginning of the FBC. Among the lessons to be drawn was the importance of the wisdom of the slogan, 'Militarize the civil and civilize the military', which meant for Biddulph closer cooperation and closer combination.[26] Biddulph argued that the parochial (in other words, colonial) view in India regarding the railways and transport more generally was a continuing obstacle to the railways remaining a civil department and not a war department. During war, such as the ongoing war India was in, there was required a 'true appreciation of [the] railway effort required in its wider aspects'. This true appreciation, Biddulph argued, could not be made based on the limited information available to a civil railway department. The QMG and the LofC were competing in the same market for the same resources and as he had personally seen, each new military exigency, including Iraq and Burma, antagonized the civil railway department and affected 'internal railway efficiency' in India. The solution, Biddulph argued, was to combine both under a single authority. The combination of all transport under central control had by this time become almost complete. Everything from roads and ports to IWT and the railways were 'virtually' in one department and thus under central control.[27]

Despite the need for central control, Biddulph also recommended specifically for the railways the decentralization of authority. Local leadership and authority needed to be encouraged to enable quick responses to immediate conditions, a lesson that had been demonstrated so clearly in the FBC, as discussed in Chapter 5. Biddulph saw a lingering parochial view of the railway's effort in the war, an isolationist outlook and insufficient delegation of authority in the railways in India as a continuing danger for the war effort. This situation, Biddulph argued, was the result of political depredation by India's most influential public utility organization. Its ethos was governed by intense competition with the roads and river transport and hence saw 'business principles as the only road to survival'. There needed to be a new ethos and this was the acceptance that winning the war had to be the only consideration. Although the war effort included to his mind 'the welfare of the civil community', every government department had to be treated as a war department and all trading had to be for the national good and not for private gain. In other words, money had to be viewed as a means and not as an end. As far as transport was concerned, Biddulph viewed this as a total war.[28]

There were also operational reasons to militarize the railwaymen themselves. As the FBC demonstrated, Biddulph argued, transport personnel in India had to be militarized. In terms of enemy action's weight ratio, the greater portion was directed against the fighting troops, but they are equipped and organized to respond to this and are aided by the possibilities of concealment and manoeuvre. The least weight is directed against depots, but base depot positions are unknown to the enemy at the beginning of the campaign; they have the benefit of camouflage, concealment and AA protection, and they generally have only to fear air attack and not infiltration on the ground. In-between these two extremes was the LofC. The LofC was 'fixed and exposed', the personnel are scattered and isolated and vulnerable to hostile civilian elements, train crews like river crews cannot take cover immediately very easily and such personnel operating in Forward Areas are liable to be encircled by the enemy or otherwise cut off. Modern war demanded of transport personnel 'the qualities of personal courage, discipline, individual resourcefulness and leadership'.[29] The FBC also revealed that there were also other material limitations to using civilian personnel. These included problems in feeding them, paying them or the military trusting them or believing that they could have the same 'fervour and singleness of purpose' in the defeat of the enemy as the military had. This situation was worsened when dealing with military men with whom they could not communicate, a clear reference to the Chinese Army. As a result, all transport personnel or at least all those in the Front Areas had to be soldiers.[30]

Colonial authorities, such as Dorman-Smith and Foucar, also agreed that the railways could have done more, but for the initial delays in the introduction of militarization to the BR caused by a clique of reluctant 'colonial' railwayman. One result was the failure to couple militarization with military training and Burma's railwaymen thus did not receive the military training that, critics later surmised, would have annealed these railwaymen, making them better able to adapt to working in the conditions of a war zone. Such colonial insiders thus directed criticism at poor imperial stewardship of the empire's defence and Burma's along with it.[31] The consensus by all involved then, regardless of whom they blamed, was that India needed railwaymen with military training before it was too late.

Adaptation on the north-eastern LofC after Burma

Some of the changes to railway operations would rely at first upon the adoption of metropolitan military texts. The WO version of the *Military Railways Rule Book*, for example, was reprinted in India in 1942. However, as Biddulph had learned in the FBC, this application of the text itself was insufficient. Local experience or better, experience in applying these procedures locally in a war zone, had to be added to the mix to make it work. Potentially the best men to train other railwaymen how to be military railwaymen were those survivors from the BR men who had not yet been allotted to other tasks. Biddulph had foreseen this while the FBC was still going on. In April 1942, Biddulph had advocated that the military should make contact with transport evacuees at the points of arrival in

India. He distributed certificates to BR staff (as well as to IFC personnel) so that they could use these both as a means of identifying themselves and to permit them to be able to obtain pay in India.[32]

Not all of the BR staff would succeed in reaching India nor would all survivors be useful or even available for further railway operations. Some of those BR men, for example, who had escaped then succumbed instead to malaria or other diseases in Indian hospitals. C. H. Whapham, for example, made it to India but died there of cholera he contracted along the way.[33] Some were retired or were put on leave preparatory to retirement, as was the case with Cardew who left for Australia in July 1942 for the duration of the war until he retired in 1944. Others, including Proctor and Parker, continued service with the GOB in exile at Simla or with the Indian Army Command or the South East Asia Command formed the following year. Edwards, for example, who had officiated temporarily as CRC for the BR in 1940, was made a lieutenant colonel and became an Assistant Director of Transportation at the South East Asia Command HQ and later the Deputy Director of Transportation with the rank of colonel and served as J. C. B. Wakeford's deputy there for two years.[34] From 13 July 1942 until 24 April 1945, Maj. T. Douglas Lee served as the Deputy Controller of Railway Priorities in the Priorities Branch of the War Transport Department.[35] Some BR men were also lent out to railways in other areas of India, including Bryant, who after over a year with Brewitt's men was lent to the South Indian Railway from July 1943 until January 1945, when he was returned to the BR.[36]

To continue recruiting these men into military transport service, the GOB extended the life of Brewitt's Railways Directorate (Burma) on 11 December 1942 by which time 3,000 BR staff had volunteered. As the rate of volunteers declined over the following months, the Railways Directorate (Burma) was finally terminated on 1 August 1943.[37] During its continued life in India before the later date, the purpose of this directorate was to manage any remaining affairs related to the militarization of the BR and to 'recruit Burma Railways personnel to military Railways units raised in India'.[38]

The degree to which these men were considered valuable for railway service in India is indicated in part by the continued service of Lewis, who had lost his leg operating trains in the Pyinmana area. His missing leg presented no obstacle to Lewis's continued climb up the ladder of command, as he eventually became a lieutenant colonel. He even took medical leave in 1944 without the details of his missing leg appearing in his personnel file. During this period, he was in regular contact with other railwaymen from Burma, among whom he continued to work, and all reports made on him were by Burma Auxiliary Force or BR officials. Lewis's work remained valuable enough to ignore the fact that he was disabled until 3 April 1945, when examiners of the India Office Medical Board found Lewis standing before them on only one leg and immediately called for his being relieved of duty.[39]

The BR men were directed into new railway units that would technically belong to India for the time being. After evacuation to India, the Burma Army and the Indian Army HQs decided to proceed with the formation of the original four military ROCs approved on 14 March 1942 in Burma as well as to form others, using a nucleus of BR staff as far as possible. As there were no railway units in the Burma Army, the GOB agreed that they would be treated as part of the Corps of Royal Indian Engineers but would be earmarked for Burma service once the opportunity for returning to Burma arose. Although many of the BR's survivors were still in the process of being located in India, by August their commissions granted in Burma were reconfirmed under the Corps of Royal Indian Engineers according to the original commission date in Burma, either 25 February or 14 March 1942.[40] These men would also now be given formal military and combined operational training at Jullundur to make them militarily effective. Biddulph argued that by comparison to the quality of the untested trainees available in India, the BR veterans were 'skilled personnel … [who] are most valuable material, and possessed of local knowledge'.[41] He suggested that the same might be true of those transport personnel who had escaped from Malaya and other now-occupied territories. Such men would be crucial not just for operations in India, but also for the reconquest of Burma and Malaya, where this 'local knowledge' would be invaluable.[42]

Training began at the end of July 1942 when thirty ex-BR officers reported to the Army Technical Centre at Jullundur, where they were housed in army bungalows and for the first time given proper military uniforms and badges of rank and many were now promoted to captain and other ranks. They then trained at No. 2 Transportation Training Centre (TTC). This included parades, lectures on bomb disposal and army regulations, as well as tactical exercises.[43] After training at Jullundur, the BR men would form a number of new units that were to be made up of only BR men with some additional training. The new units these men would form were given the suffix Burma – hence 165 Indian ROC (Burma), 168 Indian Railway Construction Company (RCC) (Burma), 170 Indian ROC (Burma), 172 Indian Railway Workshop Company (Burma) and 174 Indian Transportation Stores Company (Burma) – effectively indicating continuity with the BR after May 1942. To take one of these units as an example, 172 Indian RCC (Burma), under Officer Commanding, Maj. J. J. A. Ferguson, was formed at the end of August 1942 and included other BR men such as Capt. H. Soord, Capt. J. Hancock, Lt. A. N. Copeman, Lt. D. N. Kapur and Lt. E. W. G. Howes. Another of these units, 170 Indian ROC (Burma) became part of the establishment at the No. 2 TTC at Jullundur.[44]

The intention had been to keep these units exclusively ex-Burma so that they could be redeployed to Burma when the time came and assurances of this were given to the GOB. As there was an imbalance in the availability of BR staff from different trades, however, these units necessarily included some staff and some BR men were distributed among other, nominally Indian, railway units.[45] While this created administrative chaos in July 1944, as a separation of Burma staff from Indian staff appeared to be in order, in the meantime it widened the benefits and dissemination of their intense experience of operating the railways in a war zone among IR staff. The addition of these units of BR men, especially workshop and technical tradesmen, thus helped to raise the overall capabilities of the railways. The bulk of the Burma transport veterans, however, would be directed into helping to operate the railways on the Assam LofC. R. D. Cant was made commander of an RCC, (Royal) IE working on bridges on the Bengal–Assam Railway. Fitzherbert was promoted to major in November 1942 at Jullundur and then became 2nd in command of the No. 5 Railway Operating Group deployed on the Assam LofC (as well as in Southern India).[46]

Some BR men were posted to command positions in other Indian railway units as well. Blakeney was given command of an operating unit of the IE in February 1943 and early in 1945 was promoted to lieutenant colonel and as commander of the No. 8 Railway Operating Group. Binnie, who had joined Brewitt's group in March 1942 when many of the remainder of the BR men were militarized, was promoted to major after the evacuation and placed in command of an ROC. C. H. Morgan, promoted to major of the IE, was placed in charge of an RCC before he was made a Planning Officer with the Indian Supply Department assigned to the IR Board's Wagon Building Programme. E. J. R. Markwick, a BR man who had actually been mobilized as a British, Other Ranks to serve as Assistant Garrison Engineer when Rangoon was evacuated, joined the IE after trekking out to India and was made a major in command of a Railway Construction & Maintenance Group (RCMG). N. McAllister was also given an RCC to command and during the retaking of Burma was sent back as a lieutenant colonel with the Transportation Directorate.[47]

The collapse of Burma provided more than veteran railwaymen to India; it also provided experiences that could be taught at the new TTCs. This was particularly true of the No 2. TTC at Jullundur, which focused on military railway training. Brewitt, who, as Director of the Railways under Biddulph in Burma, had received a crash course of sorts in Longmoor training in the field, was now asked to take on the role of Chief Instructor at HQ, No. 2 TTC, a post he would fill until late 1944 when he was returned to service in an establishment earmarked for Burma.[48] In Burma, Brewitt and Biddulph brought together local, colonial knowledge, on the one hand, and imperial railway training, on the other. In the FBC, they together forged a synthesis in the course of railway operations in the midst of a Japanese invasion that kept the railways going until after the army had given up and was in retreat. Brewitt's posting was in part an admission that imperial training was not enough in dealing with the challenges of railway operations in war in a colonial setting. Brewitt's ability and willingness to apply this experience to teaching in No. 2 TTC was at least in part an admission that local experience was not enough and that this kind of formal training, based on Longmoor, was valuable and necessary.

Brewitt's presence at Jullundur was strengthened by the participation of others at the School who had worked with him during the FBC. Among them, Prendergast, who had led his No. 3 RCMG from Rangoon to Mandalay and then on to Myitkyina, doing as much for the railways as the BR men themselves, now shared his experiences with trainees at the No. 2 training centre within a few months of his return to India. Among the many sources from participants used in the present book to understand the railways' experience in this campaign are Prendergast's notes from his lectures made at the Centre on 17 July and 7 August 1942.[49]

These railwaymen and experiences from Burma were directed at building up the railway resources of the northeast, in Assam. As India saw its historic vulnerability on its northwestern frontier, its roads and railways were concentrated here and the northeastern frontier was serviced by the Bengal and Assam Railway, what was considered a secondary line only. Now that this was the area most directly threatened with invasion, India's railway build-up during the war was mainly in this area. In addition, as new military units were raised, they were sent here.[50]

Ensuring that the long, straggling line was equipped with a communications system was one of the first improvements undertaken after the close of the Burma Campaign. At the time of the close of the campaign, the Bengal and Assam Railway still had no railway control system in place. In 1943, a railway control system was put in place to rectify this. However, India was suffering very severely from the shortage of materials available because of the war. They thus had to rig together a system using second-hand wires and circuits that failed constantly. Nevertheless, railwaymen described the improvements in railway control, especially the use of telephone to communicate between stations, as a significant aid, greatly facilitating train working. Existing Indian railway train clerks manned the control offices, but Indian Operating Companies reinforced their work. The company sergeant major was usually the chief controller as well.[51] Another change was the provision of adequate air cover and AA protection. One such unit based at Gauhati was a Heavy AA Regiment, a Territorial Army unit brought in from Northern Ireland. A major advantage Assam had that Burma did not once the Japanese had begun to take large parts of Burma was that while the Japanese Army was within striking distance of Assam, their air bases were too far to the rear to launch major air raids on the railway line.[52]

The BR's experience in the FBC also affected the structure of railway management. One of the most significant changes to decentralize authority on the Bengal and Assam Railways required changing entirely the existing system of railway management on the line. The government-owned Bengal and Assam Railways had begun the war under a departmental system of management. This concentrated all authority into the hands of department heads in Calcutta. Thus, in 1942, the new military units conducted work on the lines under the civilian District Officers. As the privately owned Assam Bengal Railway's operating contract expired in January 1942 the IR Board had the opportunity to amalgamate it with the state-operated East Bengal Railway and it owned the north bank line. All lines in Assam were now operated by the same administration and they were too lengthy and unwieldy to manage from a single office anyway, so this provided the opportunity for reorganization. Thus, in 1943, the Bengal and Assam Railways' management structure was changed to a divisional organization. This decentralized authority into the hands of Divisional Superintendents, with a Deputy General Manager based at Gauhati and other stations, where they were only responsible for railway operating. Senior railwaymen like Mr Hussain, the new Deputy Manager in 1942, ran these railway centres as 'semi-independent administration[s]'.[53]

At the conclusion of the FBC, the carrying capacity of the railways in Assam was very low and poorly equipped because of colonial underinvestment.[54] The Quebec Conference in 1943, however, put greater demands on the performance on the railway in Assam. Allied leaders agreed on yearly increases in the capacity of the Assam LofC that were mainly placed upon the railway. To meet the increased tonnage required capacity had to be expanded. The colonial solution was partly influenced by the lessons learned in Burma. As Prendergast had argued and others agreed, the BR was limited in large part because of its technical aspect. This technical aspect had been conditioned by the railways in Assam having been built mainly to barely cover need, which was connected to the tea estates and the oil products from Digboi. The metre gauge line was low capacity. The coal available in the area was poorly suited to locomotives. Block stations were often fifteen miles apart from each other. It was connected to Calcutta by a broad gauge railway requiring all wagons to be unloaded and then loaded onto new wagons to make the entire journeys. The un-bridged Brahmaputra River was a major bottleneck as the loading and unloading of ferries seriously slowed traffic movement and narrowed capacity forward. Further, in 1941, a number of metre gauge locomotives and rolling stock from Assam had been sent off to Iraq.[55]

Increasing the capacity of the railways in Assam would require changing that technical aspect, improving the lines so that they were not limited to the levels required by the pre-war colonial economy. From the Indian Army's point of view, the limited capacity was due to the weakness of the line, again its technical aspect. In December 1941, the rail limitations in Assam were 600 to 1,000 tonnes per day.[56] Changes were made in the arrangement of construction engineering on the line. Such military construction projects were conducted on the line in 1942 under the Commanding Officer of a HQ, Indian RCMG under the Director of Transportation India, keeping them from train operating. In 1943, a Construction Department was created under a Chief Engineer Construction, which allowed the military railway units to concentrate on railway operating.[57]

However, GHQ India complained that it lacked a policy direction from the WO until the middle of 1943. India thus sought to meet prospective demands as it saw them on the horizon, estimating ultimate demands to be about 10,000 tonnes per day from Calcutta to Assam, not only by increasing capacity in expanding actual railway infrastructure, but also in applying Movement Control, which is beyond the scope of the present volume. The first effort was to construct railhead facilities at Manipur Road Station, the direction to which evacuees from Burma were headed in April and May 1942. The absence of either clearage facilities at Manipur Road Station or overall control over abnormal traffic brought to the military's attention the inadequacy of the railway and the need for a regulating station that led to the construction of additional sidings at Lumding. The doubling of the metre gauge line between Parbatipur and Manipur Road was recommended but at first not approved. This left improvements at the time limited to inserting additional passing stations and constructing new sidings, as well as the rail layout of a new advanced base in virgin jungle south of Manipur Road, involving a total of eighteen miles of metre-gauge track. The project was undertaken under military direction, using military labour and was finished in April 1943 despite the onslaught of malaria that affected fifty per cent of the labourers. The same month, a planning paper for the additional expansion of the railways on the Assam LofC was approved. The plans included the construction of a single line bridge over the Brahmaputra River (the first such bridge), doubling 662 miles of metre gauge line, the construction of more locomotive sheds, the addition of more transhipment lines at Parbatipur and a new regulating station for the American Airfield Area. The Americans also built a new railhead at Lekhapani to serve the Ledo Road. Although approved, the planning paper would not be implemented until August 1943, when policy direction came from the WO after the Quebec Conference, the planning paper serving as a basis upon which to identify new works.[58]

The Quebec Conference agreed that the Assam LofC was to increase its overall capacity for military stores to 4,400 tonnes a day or 132,000 tonnes per month by October 1944 (after allowing for construction materials) and to 7,300 tonnes a day or 220,000 tonnes per month by January 1946.[59] As far as the Indian Army was concerned, increases would have to depend upon new construction. The first increase would be achieved in part by doubling the broad gauge section of the line between Abdulpur and Santahar, the metre gauge section between Parbatipur and Golakganj, the metre gauge section between Lumding and Dimapur, the conversion of the 2.5-foot railway line between Jorhat and Mariani to metre gauge and the construction of a railhead for the Chabua area. These changes would also require more metre gauge locomotives and rolling stock. The second increase would include the doubling of the broad gauge railway between Santahar and Parbatipur, the metre gauge section between Golakganj and Amingaon, the metre gauge section between Pandu and Lumding, the metre gauge section between Dimapur and Tinsukia, and development of the line between Chandranathpur and the hill section of Lumding to bring up 500 tonnes of stores loaded at the port of Chittagong. There were also other improvements to railway infrastructure, including a regulating station at Mariani, new terminal developments and a base layout at Parbatipur to catch traffic between the broad and metre gauge sections here and thus direct a more even flow of railway traffic forward to Dimapur.[60]

Experiences in railway security were also carried from Burma onto the railways of the Assam LofC. Goddard was put in charge of the LofC troops in

Assam on his exit from Burma. Some of the men who played various other roles related to the railway and railway security were also put in place on the railways in Assam and undertook the same local measures they had learned had been so effective in Burma. One of these officers was Tony Mains, who was now locally responsible for security at Gauhati and other LofC centres. While the railways in Assam would not be subject to the same intensive operating conditions of the BR as they were being overrun, security and sabotage were claimed to have been widespread in Assam. Certainly some of this was exaggerated. Mains relates one case of reported sabotage that he investigated, in which saboteurs were claimed to have blown up a shunting engine in Pandu Railway Yard and killed the crew. As other crews believed explosives were mixed into the coal they refused to work, stopping rail movement. On investigating, Mains found that the crew appeared to have mistaken a can of blasting powder for canned food and when they tried to open it on the footplate, it exploded. Again, as in Burma, the potential for a repeat of many of the fears that worked against the efficient operations of the railway was also present in Assam.[61]

On the other hand, at times, operating conditions for the railways in Assam did include actual threats from the local population as in the rebellion of 1942. There were episodes of derailments that were clearly the work of sabotage, although it remained unclear how much was connected to the Japanese and how much to local resentment of colonial rule. The fact that there was believed to be genuine sabotage at work brought some of the lessons from the Burma Campaign to bear once again on Assam's railway operations. One of these was that the line and other important railway installations had to be protected in the context of an inadequate supply of actual soldiers or railway staff would be unwilling to work. One solution put in place on the railways on the Assam LofC in late 1942 included Mains. This was a three-tiered system of security. At the most local level, the Assam Government issued orders that required local villages to patrol in village protection parties the parts of the line to which they were adjacent and if they did not, collective fines would be imposed. Second, Ghurkhas domiciled in the province were recruited by the Inspector General of Police to form a 1,000-man strong Auxiliary Police codenamed Railforce, armed with a rifle and five rounds of ammunition, that would be based at stations in the most dangerous areas and in their patrols bolster and monitor the work of the village parties. Third, an Indian Territorial Battalion codenamed Railtroops was deployed, headquartered at Lumding with detachments at other important railway centres such as at Gauhati. The working of all three tiers of the scheme was the responsibility of the commander of this battalion.[62]

Imperial flexibility

Lessons learned in Burma were of relevance only to South Asia and other 'Eastern countries', at first. As mentioned, the WO version of the *Military Railways Rule Book* was reprinted in India in 1942. This was followed in 1944 by the publication of a new *Military Railways Rule Book* entirely directed at application locally, in India, Burma and Malaya (in advance of its impending reconquest), as indicated in the title, *Military Railways: Rule book for Use by Indian Railway Troops in India and Other Eastern Countries*. While the general layout and many of the rules themselves remain the same, interspersed are new rules not found in the original that provided flexibility to local operating superintendents to deal with local variations in the local technical aspect of railway sections. For example, in the section entitled 'Block Working Regulations', the 1938 original begins with the topic 'Cardinal Principles of Absolute Block Working'. The 1944 edition includes the same principles, verbatim. However, it precedes them with an entirely new section, 'Alternative methods of block working', which lays out the full range of methods of absolute block working that might be employed by the order of the operating superintendent, as well as three additional methods of working in emergencies, including 'time interval', 'pilot guard', and 'permissive'.[63] Such inserts of additional material are found throughout, making the 1944 edition a localized version.

Flexibility in localizing military railway operating to local contexts meant that there would not be one manual for the empire. Nevertheless, it did mean that in this way, metropolitan rules could be simultaneously both imperially valid and localized. What had been military railway operating designed for north-western Europe now became suitable for extra-European application, something that would become increasingly important in the decades ahead.

Additionally, the increased currency of railway operating procedure had another significant result. Organizationally, India's Transportation organization edged closer from late 1943 to uniformity with Transportation and Movement Control across the empire. Lt. Gen. A. R. Godwin-Austen, who was made QMG in India in late 1943, was committed to evolving Transportation and Movement Control into an imperial organization. Experiences in Italy had shown that it was important to clearly demarcate Staff and Service functions regarding Transportation and Movements and the benefits of keeping them under the technical control of their service head. It was also important that as Transportation Units, Transportation Officers and Movements Officers were being interchanged between Theatres that they be 'accustomed to the same organization and channels of control'.[64]

Transportation and Movement Control would continue to move towards an imperial organizational and technical approach to transport. This would include the combination in India in 1944 of Transportation and Movements as Biddulph had urged. Although increasingly standardized and imperial, the Transportation Technique still allowed for cooperation of local operating staff, the importance of which Burma had demonstrated.

Chapter conclusion

After the end of the FBC, most parties directed the blame for the failures of the campaign at almost anyone else. Cooler, less personal assessments would take years to vindicate many of those blamed by one or another report, letter or account in the remaining years of the war. The management of the railways and what or who accounted for its failures was the subject of fierce debate. Despite criticism of the Transportation Directorate, the commitment to a stronger military Transportation and Movement Control organization in India was one outcome of the FBC. This was the result of campaigning by the reformers and 'prophets' of military Transportation. Adaptation in the case of Transportation Technique had been a long and difficult process in part because of opposition rather than aid from the top of particular branches of the Indian Army. There would be additional developments on the Assam LofC in 1944, most notably the arrival of American railroaders. This would lead to a recasting of railway operating developments and successes in Assam as an American achievement, much to the chagrin of British and colonial railwaymen and engineers who had been hard at work in the theatre for several years and who continued to run much of the railways with success. This development is covered elsewhere in the context of Movement Control.[65]

The BR men who wed military Transportation together with local colonial knowledge and experience in wartime put the Technique into practice in military railway operations. This fully transformed what had been a metropolitan Technique into an imperial one and the process, by which it had been introduced, carrying Biddulph in particular to various fronts of the war and across the empire, had proven to be imperial as well by necessity. The impact of imperial Transportation (and Movement Control) proved durable and, in the hands of other Transportation and Movements officers elsewhere in the empire, on different battlefronts, developed into an all-purpose military method that would aid global, British deployments in later generations.

Conclusion

However vital they were to the defence of and evacuation from Burma, the railways did not emerge from the conflict as the success story that they appear to have been. This was partly due to the early intransigence of Indian military authorities who did not see the necessity or plausibility of militarizing transport in the subcontinent. For all the criticism he received as a commander, Field Marshal Sir Douglas Haig was open to all kinds of technical innovations on the Western Front in the First World War, not just the tank but also, especially relevant here, the reworking of railway logistical support by Eric Geddes.[1] Transportation Technique was not received with the same open arms.

Biddulph played a crucial role in interpreting the results of the Burma Campaign for transport authorities in India. As mentioned in Chapter 3 and discussed in Chapter 8, Biddulph had argued the case that the railways' success in Burma was only possible because of the continuous military reverses that created a backward flow of essential manpower, equipment and resources. The railways remained, to the end of the campaign, without the potential to support a successful military defence. They certainly would have proved inadequate for a successful counteroffensive against the Japanese. There were huge internal problems including procedural failures and mismanagement in keeping mass transport going until Japanese forces had won the campaign. Reforms in military Transportation in India had to be undertaken quickly and radically, if India were to survive an invasion by the Japanese and if Britain were to win the war in Asia. This would only occur if the railways' continued survival was underplayed and a defeat paradigm, wherein all elements of the defence were brushed in the same colour of failure, was fully allowed to play its part by serving as a necessary catalyst for reform.

One of the reasons why the scholarship on the role of the Transportation experience of the FBC and its role as a catalyst for later military Transportation changes in India is so thin is that some of the key sources have remained until now out of reach. Much crucial documentation remained in Biddulph's hands and would not be accessed directly by anyone else until 2015 when the Biddulph family permitted me access to the papers in Biddulph's suitcase. Biddulph had lent Dorman-Smith his own report on the FBC and had penned in corrections on Kirby's chapters before publication. However, the reports by railwaymen during the campaign, his own correspondence, diary, handbook and other miscellaneous records remained out of reach by the series of scholars who scrutinized the FBC for its secrets, but found little to say about the railways.[2] The Biddulph Papers and other archival materials used for the first time in the present volume allow the most complete assessment of the BR's performance in the FBC, its failures, its successes and its impact than has thus far been possible.

Brigadier R. Micklem, who penned the War Office's official history of Transportation during the Second World War in 1950, had much to say about the organization of Transportation Directorates in various theatres, countries and colonies, but little to say about the movement of Transportation Technique from one operational area to the next or the agents who carried it. Indeed, as Micklem correctly points out, after the FBC and related defeats in the period, colonies as far away as Nigeria immediately adopted Transportation Technique, set up Transportation Directorates and formed military ROCs. However, these are presented as reforms that simply happened without carriers and without contest.[3]

If successful, innovation and adaptation are endpoints to a single process, however twisting and frayed that process might be. Innovators develop a new or better idea or technology and adapters put it into play, as Albert Palazzo has shown so well.[4] Between these two points is another, essential group. I hesitate to label these men innovators for innovation involved a great many people many of whom never pursued or at least often never achieved change. Instead, these men were a more select group who were committed, often passionately, to teaching others a better way of doing things. Biddulph called himself a prophet. Again, he was not, for he was no blind carrier who felt his subject was immune to improvement, but he was still a reformer. There are also often invisible men and women. Most reformers are left out of the history of technology and military adaptation in large part because their work is not so well documented, carried out as it often is in harsh political infighting behind closed doors, in boardrooms, at army headquarters and the like. Innovators and adaptors, by contrast, are the sung heroes of military histories. The dirt and mud that must be slogged to carry reform to adaptation is often such that it guarantees reformers an anonymity that obscures the crucial historical role they play. The obvious exception would be high-level politicians, such as Viscount Richard Burdon Haldane, Secretary of State for War until 1915, who suffer in the public eye.[5] As the experiences of attempting to introduce Transportation Technique into India and Burma have shown, military adaptation could not have occurred without innovators who could find opportunities to introduce change and reform. This book has looked closely at Biddulph, but there were Royal Engineers in various other theatres who also played significant roles. A close examination of their efforts in the war indicates that without their agency in the transitioning of Longmoor training into largely extra-European world application through further field development and promotion, the lessons of the FBC might not have yielded their benefits for Transportation Technique.

To examine the role played by reformers requires mobilizing biography as a historical method because their campaigns are like those of conventional war fought over time and space, in different venues, with failures and successes, sometimes leading to victory but often to defeat. It is important to understand not only the question of who and what, but also where, when and why. Other historical methods are necessary for uncovering how, the most important question, and this is often found not in the official reports and histories, but in the quiet truths recorded in private correspondence or in personal diaries, the latter often intended only for private contemplation. All of these materials, including the documentation of the innovations and adaptations, are necessary for arriving at a fuller picture of why reforms occurred and who was responsible.

Reform can be dangerous business. Changing knowledge was politically threatening to powerful elites. Reformers thus found it necessary to break down institutional structures and reform them with select personnel. This meant recruiting allies and relying upon them in the new system. When things soured for whatever reason, they shouldered the blame personally. Biddulph suffered a major hiccup in his career, however temporary. Biddulph was kept from the Transportation Directorship of India, something he seems to have resented, however much he was vindicated and given a further position of confidence and authority relative to transport later on. As a process that depended very heavily on personal agency, the success or failure of military adaptation was influenced to a strong degree by the effect of personality. Things might have gone somewhat more smoothly in both India and Burma, for example, but for an irony. Biddulph, who spent his career teaching military men and whom Vickers later described as not being the best at working with civilians, found himself trying to convince civilian transport people as well as military men of the value of military

Transportation Technique and only really succeeded where he could militarize the civilian staff and run things in a military way. Manton who had successfully made the transition in Britain from military to civilian railway training was sent to run military Transportation in India where the challenge was to win over military men. Had their positions been reversed one might imagine an easier time having been experienced by these two men in Burma and in India in late 1941 and 1942, respectively. As bad as relations with the governor became, these could have been much worse if Dorman-Smith had not come to Burma in 1941 predisposed to the need to prepare Burma for possible war.

More broadly, this book has been about military adaptation in the context of the imperial circulation of knowledge. The newly available and otherwise untapped archival sources used in the present study also allow the most complete picture, again until now, of the difficulties encountered in introducing Transportation Technique to the Empire. Models of imperial circulation of knowledge tend to focus on movements of ideas and technology within knowledge communities stretched across the empire or of culture across societies within the empire. Crossover knowledge involved more struggle and more pain because such communities represented particular clusters of political relationships built on control of knowledge and thus the community. As a technology that required winning over both military and civilian communities connected to transport and supply, Transportation Technique's prophets were bound to face troubles. This was clearly experienced in the FBC and such communities had to be ripped apart.

As this book has attempted to show, a long-term, entrenched, European elite, who ran the railways in colonial Burma, had a complicated relationship with imperial military Transportation authorities. On one level, the problem was not unique to the colonial world but represented the kinds of issues that set central and local authorities against each other in all kinds of institutions, not just the railways. In the present book, the railwaymen's claims that Biddulph and others did poorly because they were *outsiders* and knew nothing about Burma, for example, would have likely been voiced no matter how the campaign had gone. The metropolitan military authority's lack of the *local* railwayman's peculiar knowledge regarding the *local* railways was their rationale for this resistance but it was the threat to the *local* railwaymen's peculiar authority which rested upon this knowledge that was the problem, not the course of the campaign. *Local* railway officers were reluctant to surrender the institutional power they had amassed within the railway administration to *external* (i.e. imperial) authorities, civilian or military. In their eyes, to do so would have meant abandoning the fruits of building a career (and a powerbase) within an institution, whether government or commercial and whether in the West, Burma or anywhere else in the imperial world, this behaviour is not surprising.

On another level, the conflict between local railwaymen and metropolitan military railwaymen was entirely a late colonial phenomenon. What made European expatriate disinclination to accept metropolitan military innovation powerful in the FBC was the threat it posed not only to the former's authority but also to their occupational position in the colony altogether. Constitutional developments in the 1930s in response to nationalism and other factors erected increasingly difficult obstacles for imperial interests to cross and reluctance by different parts of the empire to sacrifice their own resources on behalf of others. The problems facing the BR in 1941 to 1942 were symptomatic of an empire that was already unravelling. White colonial railwaymen clung assiduously to their authority on the basis of local experience as their position on the colonial racial hierarchy was rapidly becoming unimportant as the Burmese expanded their political position in the colony and Burmanization was making itself felt throughout the colonial administration.

The meeting of imperial and colonial interests, approaches and men thus produced a tension that created a problem for railway operations perhaps greater than any other major factor. This friction between imperial and colonial railway authorities was resolved by undoing the railways as a colonial institution and freeing up the various resources, both material and human, to reorganize as best they could to operate on their own, but with their morale bolstered by militarization and their manpower supplemented by small contingents of military engineers. The BR failed miserably until Biddulph took what turned out to be the wise step of decentralizing authority to small groups of men with local, colonial expertise regarding the railway system they were operating who could keep the local sections in operation. While reports on the campaign highlighted the failures of the railway as viewed through the decline of central operations, the evidence from the local sections reveals the ability of the local railway managers to operate autonomously in support of demands on the railway at the local level. The BR men kept the BR functioning to the last breath of the defence of the colony.

Transportation Technique's eventual successful adoption in India and its concomitant transition to an imperial approach required this ordeal by fire and sword to test it, to temper it and to hasten its digestion in the FBC's aftermath. Ultimately, the Transportation Technique that came out of the fighting in Burma, what Biddulph described in the midst of the evacuation of Rangoon as a holocaust, was not a different creature than the technique taught at Longmoor, but it was a stronger one and it could handle a larger world. The military Transportation that emerged out of the FBC was most directly significant for the Assam lines of communication and the emergence of modern military Transportation in India. But the Technique would continue to develop, in combination with Movement Control, and would become a technique that was globally applicable, something it was not in 1941, but something that strengthens the British Army today and has since the Second World War in a wide variety of international deployments of greater and lesser scale, a Technique for all seasons.

Notes

Introduction

1 See, for example, Michael D. Leigh, *The Evacuation of Civilians from Burma: Analysing the 1942 Colonial Disaster* (London: Bloomsbury, 2014); Philip Woods, *Reporting Retreat: War Correspondents in Burma* (London: Hurst, 2017). Another work on the role of the civil administration of Singapore in 1941–2 has reaffirmed the importance of looking at additional, ignored aspects of these campaigns. See Ronald McCrum, *The Men Who Lost Singapore* (Singapore: Singapore University Press, 2017).

2 'Burma-Siam Railway Policy', *The Singapore Free Press and Mercantile Advertiser* (5 September 1930), 12.

3 Stephen L. Keck, *British Burma in the New Century, 1895–1918* (Houndmills: Palgrave Macmillan, 2015).

4 The 'Webs of empire' model reflecting the uneven and horizontal movement of ideas in the imperial circulation of knowledge was introduced in Tony Ballantyne, *Orientalism and Race: Aryanism in the British Empire* (Cambridge: Cambridge University Press, 2002).

5 J. S. Furnivall, *Colonial Policy and Practice: A Comparative Study of Burma and Netherlands India* (New York: New York University Press, 1948), 304–5.

6 Pradeep Barua, *The Army Officer Corps and Military Modernisation in Later Colonial India* (Hull: University of Hull Press, 1999), 126–7.

7 Jackson, *Distant Drums: The Role of Colonies in British Imperial Warfare*, 251–61.

8 Ibid., 250.

9 In 1941, the BR had 358 locomotives in operation. Myint Swei, 'Myanma-pyi Mi-yahta Hnit-taya Thamaing', *Thwei-thauk Magazine* 318 (June, 1972), 148; Francis John Biddulph, 'Report on the Burma Railways in Special Reference to Hostilities in Burma', 1 November 1942, 2, WO/203/5722, PRO; 'The Burma Railways during Invasion', *Railway Magazine* 90 (July and August, 1944), 216.

10 On occupational status groups, see Max Weber, *Economy and Society: An Outline of Interpretive Sociology* (Berkeley: University of California Press, 1978), 305–7.

11 Andrew Goss, *The Floracrats: State-Sponsored Science and the Failure of the Enlightenment in Indonesia* (Madison: The University of Wisconsin Press, 2011), 7.

12 Norman Davis, 'An Information-based Revolution in Military Affairs', in *In Athena's Camp: Preparing for Conflict in the Information Age*, ed. John Arquilla and David Ronfeldt (Santa Monica, CA: RAND, 1997), 90.

13 Daniel P. Marston, *Phoenix from the Ashes: The Indian Army in the Burma Campaign* (Westport, CT: Greenwood, 2003), 79–99.
14 Tim Moreman, *The Jungle, the Japanese and the British Commonwealth Armies at War, 1941–1945* (London: Frank Cass, 2005), 97–105.
15 Graham Dunlop, *Military Economics, Culture, and Logistics in the Burma Campaign, 1942–1945* (London: Pickering & Chatto, 2009), 7, 10.
16 Julian Thompson, *Lifeblood of War: Logistics Armed Conflict* (London: Brassey's, 1991), 82.
17 J. P. Harris, *Men, Ideas, and Tanks: British Military Thought and Armoured Forces, 1903–1939* (Manchester: Manchester University Press, 1995); Albert Palazzo. *Seeking Victory on the Western Front: The British Army and Chemical Warfare in World War I* (Lincoln: University of Nebraska Press, 2000).
18 Mark Harrison, *Medicine & Victory: British Military Medicine in the Second World War* (Oxford: Oxford University Press, 2004).

1 Metropolitan Transportation Technique from Britain to India and Iraq

1 G. Williams, *Citizen Soldiers of the Royal Engineers Transportation and Movements and the Royal Army Service Corps, 1859–1965* (London: Institute of the Royal Corps of Transport, 1969), 87.
2 On Thompson, see B. L. Raina (ed.), *Campaigns in the Eastern Theatre*, Official History of the Indian Armed Forces in the Second World War 1939–45. Medical Services (New Delhi: Combined Inter-Services Historical Section, India & Pakistan, 1964), 124.
3 Vic Mitchell & Keith Smith, *Branch Lines to Longmoor* (Midhurst: Middleton Press, 1987), 2; J. B. Higham & E. A. Knighton (comps.), *The Second World War 1939–1945 – Army – Movements* (London: WO, 1955), 11–12; Hampden Gordon, *The War Office* (London: Putnam, 1935), 156.
4 Gordon, *The War Office*, 156.
5 This paragraph depends heavily on the official history of movements in the Second World War, Higham & Knighton, *Movements*, 3–4.
6 Francis John Biddulph, 'Signal School at the Railway Training Centre, Royal Engineers, Longmoor', *The Railway Gazette* (2 April 1937), 656–61.
7 Ibid.
8 James Hevia, *The Imperial Security State: British Colonial Knowledge and Empire-Building in Asia* (Cambridge: Cambridge University Press, 2012), 221.
9 F. W. Perry, *The Commonwealth Armies: Manpower and Organisation in Two World Wars* (Manchester: Manchester University Press, 1988), 43.

10 Ibid., 42, 44.
11 Tim Strangleman, 'Railway and Grade: the Historical Construction of Contemporary Identities' (PhD diss., University of Durham, Durham, 1998), 18.
12 WO, *Notes on Military Railway Engineering Part IV: Operating* (London: The QMG, 1942), 88.
13 For example, the first volume of the military railway engineering manuals, the one devoted to a survey, reprinted in its appendices a lengthy extract from the *Rules for the Preparation of Railway Projects* as published by the Government of India (GOI) in 1926. WO, *Notes on Military Railway Engineering Part I (Survey)* (London: The QMG, 1940), 68–77.
14 WO, *Notes on Military Railway Engineering Part II Engineering* (London: The QMG, 1940), 167–75.
15 This paragraph depends heavily on the official history of movements in the Second World War, Higham & Knighton, *Movements*, 3–4.
16 R. D. Spierts, 'Longmoor Training Course Lecture Notes', 1944, 7, unpublished ms. in private possession.
17 Higham & Knighton, *Movements*, 11–12.
18 Ibid., 7.
19 Williams, *Citizen Soldiers*, 82.
20 David W. Ronald & Mike Christensen, *The Longmoor Military Railway: A New History*, vol. 1, *1903–1909* (Lydney: Lightmoor Press, 2012), 113.
21 *Locomotive, Railway Carriage and Wagon Review* (1937), 123; *Railway Gazette International* 96 (1952), 47; Chris Manton, Lionel Manton's grandson, personal communication, 14 November 2014.
22 Lionel Manton, 'Communication', *Journal of the Institute of Transport* 19 (1938), 285.
23 R. Bell, *History of the British Railways During the War, 1939–45* (London: Railway Gazette, 1946), 56.
24 Ibid.
25 Higham & Knighton, *Movements*, 12; Ronald & Christensen, *The Longmoor Military Railway*: Vol. 1, 114.
26 This brief survey has been culled from various years of the War Department's Army List.
27 Francis John Biddulph, 'Military Transportation in India', 25 August 1942, 2–3, File 4, BFP, London, UK.
28 E. W. C. Sandes, *The Indian Sappers & Miners* (Kirkee: Institution of Royal Engineers, 1948), 669–91.
29 Biddulph, 'Report on Transportation Liaison Visit to the Middle East', 18 April 1941, 1, File 4, BFP.
30 Troopers to Army, India, New Delhi, 19 April 1941, File 4, BFP.

31 Biddulph to Langley, 19 May 194[1], 1, File 4, BFP.
32 Troopers to Army, India, 8 May 1941, File 4, BFP; Biddulph, 'Military Transportation in India', 1–3.
33 Biddulph to Langley, 19 May 194[1], 1; Troopers to Army, India, 29 May 1941, File 4, BFP
34 Biddulph to Langley, 19 May 194[1], 1.
35 Thomas R. Metcalf, *Imperial Connections: India in the Indian Ocean Arena, 1860–1920* (Berkeley: University of California Press, 2007).
36 Ronald & Christensen, *The Longmoor Military Railway*: Vol. 1, 114.
37 Biddulph, 'Military Transportation in India' 3–4, 8.
38 Ibid., 8.
39 Ibid., 3.
40 Biddulph to Langley, 19 May 194[1], 1.
41 S. Verma & Vijay Kumar Anand, *The Corps of Indian Engineers, 1939–1947* (Delhi: Historical Section, Ministry of Defence, GOI, 1974), 245; Biddulph, 'Military Transportation in India', 1.
42 Biddulph, 'Military Transportation in India', 1, 4; Verma & Anand, *The Corps of Indian Engineers, 1939–1947*, 245.
43 Biddulph to McMullen, 1 July 1941, 1, File 4, BFP; Biddulph, 'Military Transportation in India', 1, 3.
44 Biddulph, 'Military Transportation in India', 3, 6–7.
45 Biddulph to McMullen, 1 July 1941, 1.
46 Ibid., 8.
47 Ibid.; Biddulph to McMullen, 16 September 1941, 1, File 4, BFP; Biddulph, 'Memo on Transportation in India', 2, File 4, BFP.
48 R. Micklem (comp.), *The Second World War 1939–1945 Army: Transportation* (London: WO, 1950), 181; Biddulph, 'Memo on Transportation in India', 2.
49 Micklem, *Transportation*, 181–2; Verma & Anand, *The Corps of Indian Engineers, 1939–1947*, 245.
50 William Hillary Prendergast obituary in *Transactions of the Civil Engineers of Ireland* 84 (1958), 232; *Supplement to the London Gazette* (23 April 1942), 1796; 'Dictionary of Irish Architects 1720–1940', http://www.dia.le/architects/view/4411 (accessed 28 November 2014).
51 Ibid.
52 War Diary of the No. 3 RCMG, for January to July 1942, 3, WO/172/1021/3, PRO.

2 Local, colonial railway experience

1 Biddulph, 'Report on the Burma Railways', 2.

2 *The Mitre [Christ Church School Magazine of Perth]* 1, no. 7 (April, 1920), 12; *The Railway Engineering Journal* 1–2 (1972), 33.

3 James Ferguson, 'Anthropology and Its Evil Twin: 'Development' in the Constitution of a Discipline', in *International Development and the Social Sciences: Essays on the History and Politics of Knowledge*, Frederick Cooper & Randall Packard (Berkeley: University of California Press, 1997), 168.

4 See the essays in Alan Bicker, Paul Sillitoe, & John Pottier (eds.), *Essays in Development and Local Knowledge: New Approaches to Issues in Natural Resources Management, Conservation and Agriculture* (Abingdon: Psychology Press, 2004).

5 Benjamin R. Smith, '"Indigenous" and "Scientific" Knowledge in the Central Cape York Peninsula', in *Local Science vs. Global Science: Approaches to Indigenous Knowledge in International Development*, ed. Paul Sillitoe (New York: Berghahn Books, 2013), 77.

6 James C. Scott, *Seeing Like a State: How Certain Schemes to Improve the Human Condition Have Failed* (New Haven: Yale University Press, 1998), 6.

7 Ibid.

8 Aaron Moore, *Constructing East Asia: Technology, Ideology, and Empire in Japan's Wartime Era, 1931–1945* (Stanford: Stanford University Press, 2013), 137–8.

9 E. P. Anderson, 'Railway Organization in War', *Journal of the United Services Institution* 72 (1927), 510.

10 Herbert Hoover, *The Memoirs of Herbert Hoover: Years of Adventure* (New York: Macmillan, 1951), 95–6.

11 This was true of Air, Brewitt, E. V. M. Powell, George A. Cambridge, E. J. R. Markwick, Reginald Melville Ward Lowe, Thomas Clifford Parker, A. G. Van Der Beek, and others. W. J. Air Personnel File, Confidential Report, LSG/11/8/1, IOR; C. P. Brewitt Personnel File, Confidential Report, L/SG/11/8/34, IOR; Report on Mr E. V. M. Powell, LSG/11/8/210, IOR; Confidential Report on A. G. Van Der Beek, 1943, L/SG/11/8/268, IOR; Confidential Report on Thomas Clifford Parker, L/SG/11/8/194, IOR.

12 Petition of Mr P. Manifold, M/3/188, IOR.

13 Frank McKenna, *The Railway Workers 1840–1970* (London: Faber and Faber, 1980), 63.

14 Joseph A. Olmstead, *Leading Groups in Stressful Times: Teams, Work Units, and Task Forces* (Westport: Quorum Books, 2002), 57.

15 McKenna, *The Railway Workers 1840–1970*, 30–1.

16 Letter to E. Foucar, 23 July 1943, 4.

17 Report on Mr E. V. M. Powell.

18 E. Proctor Memorial, 6 May 1945, M/4/2205, IOR.

19 After the war, Air returned to university where he received his BSc in civil engineering and worked for a private engineering firm on power stations before he

was hired by the BR Home Board in 1922. *Railway Gazette International* 88 (1948), 257; W. J. Air Personnel File, LSG/11/8/1, IOR.

20 C. P. Brewitt Personnel File, LSG/11/8/34, IOR.

21 Enclosure 28, 'The Report on Burma Communications', 3, WO/106/2654, PRO.

22 Confidential report for W. R. Williams, L/SG/11/8/229, IOR.

23 *Railway Gazette* (6 January 1956), 11.

24 Ian J. Kerr, *Engines of Change: the Railroads that Made India* (Westport, CT: Greenwood, 2007), 173, n. 14.

25 Maury Klein, *The Life & Legend of E. H. Harriman* (Chapel Hill: North Carolina University Press, 2000), 277–8.

26 *Railway Gazette*, 6 January 1956, 11.

27 Ibid.; Kerr, *Engines of Change*, 173, n. 14; Klein, *The Life & Legend of E. H. Harriman*, 278.

28 *Railway Gazette*, 6 January 1956, 11.

29 Ralph L. Wedgwood, 'Paper by Ralph L. Wedgwood', *The Railway News* 84 (1905), 789.

30 Ibid.

31 *Railway Gazette*, 6 January 1956, 11.

32 Weber, *Economy and Society*, 343.

33 James Russell Andrus, *Burmese Economic Life* (Stanford: Stanford University Press, 1947), 235; John LeRoy Christian, *Burma and the Japanese Invader* (Bombay: Thacker, 1945), 130–1.

34 GOI, *Report of the Indian Railway Enquiry Committee, 1937*, 1–2.

35 Ibid., 43.

36 Ibid., 29, 45.

37 'Burma, Bold Policy of Reorganisation, 1938', 1.

38 In the new scheme for railway district organization, the number of executive districts was reduced from four to three in the Locomotive Carriage and Wagon Department and from six to five in each of the Engineering (Way and Works) and Traffic (Transportation) Departments. Ibid., 2.

39 The Secretary, BR Board, to the GOB, Department of Commerce and Industry (Commerce Branch, Rangoon), 1 June 1938, M/3/478, IOR.

40 'Notes on Chapter VIII of the Wedgwood Report, by the Traffic Manager, Burma Railways', 1–2, M/3/478, IOR; *A Record of Large-Scale Organisation and Management, 1923–1946* (London: London Midland and Scottish Railway Company, 1946), 16.

41 *A Record of Large-Scale Organisation and Management, 1923–1946*, 16; JH Wise to the Under Secretary of State for Burma, Burma Office, London, 17 June 1938, 1, M/3/478, IOR.

42 Wise to the Under Secretary of State for Burma, 17 June 1938, 1.

43 Brewitt Personnel File.
44 GOB, *Annual Report of the Burma Railway Board, 1940–41*, 33.
45 Ibid.
46 E.C.V. Foucar, 'Draft Narrative of the First Burma Campaign December 1- May 1942', E Coll 3 L.C./37 Part 1 & 2, 200; Brewitt Personnel File.
47 Minute for 30 January 1940, Burma Adviser & Services Committee, M/3/1021, 1, IOR; D. B. Petch to the Undersecretary of State for Burma, 22 December 1939, M/3/1021, 1, IOR; W. H. Durrell to Governor of Burma, 2 November 1939, M/3/1021, 1, IOR; Copy of Note, dated the 24th November 1939, from Mr W. H. Durrell, Electrical Engineer, Insein, to the Loco. & Carriage Superintendent, Burma Railways, M/3/1021, 1, IOR.
48 Copy of Note by Chief Railway Commissioner, 1.
49 J. Thomson to Durrell, 6 September 1940, M/3/1021, IOR.
50 Copy of Note by Chief Railway Commissioner, 1.

3 The failure of civilian management of a colonial railway in wartime

1 Letter to E. Foucar, 23 July 1943, 4, WO/203/4731, PRO.
2 Foucar, 'Draft Narrative', 199.
3 Telegram, WO to GOC Malaya, 7 December 1941, L/WS/1/695, IOR, British Library, London.
4 Foucar, 'Draft Narrative', 199; Letter to E. Foucar, 23 July 1943, 4; RH Dorman-Smith, 'Comments on General Hutton's Despatches', Hutton 3–12, Hutton Papers, LHCMA, King's College London.
5 'Note on Transportation Facilities in Burma for Burma Transport Co-ordination Board', 1, WO/106/2654, PRO.
6 RH Dorman-Smith, *Report on the Burma Campaign, 1941–1942*, 171, M/8/15, IOR; B. R. Pearn, 'Report on the Civil Evacuation', 5, TS/28/515, PRO; N. Johnson, 'A Review of the Burma Railways', *The Engineer* (February 1949): 178; Christopher Shores, Brian Cull & Yasuho Izawa (eds.), *Bloody Shambles*, vol. 1., *The Drift to War to the Fall of Singapore* (London: Grub Street, 1992), 247; Ikuhiko Hata, Yasuho Izawa, & Christopher Shores (eds.), *Japanese Army Fighter Aces 1931–45* (Mechanicsburg, PA: Stackpole Books, 2002), 33; Bayly & Tim Harper, *Forgotten Armies: Britain's Asian Empire and the War with Japan* (London: Penguin Books, 2004), 156–7.
7 H. L. Thompson, *The Official History of New Zealand in World War II. New Zealanders with the Royal Air Force Vol. III* (Wellington, Historical Publications Branch, 1959), 274.

8 Kin Thida Oung, *A Twentieth Century Burmese Matriarch* (n.p.: Kin Thida Oung, 2007), 90.
9 Foucar, 'Draft Narrative', 89.
10 Thompson, *The Official History of New Zealand in World War II*, 274.
11 'A Record of the 1st Bn. Gloucestershire Regiment (the 28th) in Burma, 1938–1942', 12, CAB/106/25, PRO.
12 Dorman-Smith, *Report on the Burma Campaign, 1941–1942*, 171.
13 A. A. Mains, *A Soldier with Railways* (Chippenham: Picton, 1994), 57.
14 Dorman-Smith, *Report on the Burma Campaign, 1941–1942*, 171.
15 Foucar, 'Draft Narrative', 89.
16 Leland Stowe, *They Shall Not Sleep* (New York: A.A. Knopf, 1944), 90.
17 HQ, American Army Forces China, Burma and India, 'The Campaign in Burma (from March 1st to June 1st)', 33.
18 Robert Mole, *The Temple Bells are Calling: A Personal Record of the Last Years of British Rule in Burma* (London: Pentland Books, 2001), 128.
19 Jack Cardew, 'Extracts from the Diary of Jack Cardew, Deputy Locomotive Superintendent, Burma Railways, Insein', 1, Anglo-Burmese Library http://www.ablmembersarea.com/treak-out-of-burma-military-records-html [accessed 26 April 2015].
20 Enclosure 28, 'The Report on Burma Communications', 3, WO/106/2654, PRO.
21 Foucar, 'Draft Narrative', 199.
22 Johnson, 'A Review of the Burma Railways', 178; Foucar, 'Draft Narrative', 199; Strong, 'Narrative of 2nd July 1942', 1–4, File 4, BFP
23 The evidence for this, however scanty, comes from Biddulph's personal notes penned into the margins of Strong, 'Narrative of 2nd July 1942', 3, File 4, BFP.
24 Johnson, 'A Review of the Burma Railways', 178; Foucar, 'Draft Narrative', 199; Robert Slater, *Guns through the Arcady: Burma and the Burma Road* (Madras: Diocesan Press, 1943), 220; Francis John Biddulph, 'Narrative of Military Railways in Burma', 2, File 4, BFP.
25 Letter to E. Foucar, 23 July 1943, 4.
26 Micklem, *Transportation*, 199; Biddulph, 'Narrative of Military Railways, 2, 4; Milne to Foucar, 23 March 1943, 1.
27 This paragraph relies on S. Woodburn Kirby, *The War Against Japan* (London: HMSO, 1958), vol. 2, 465; Biddulph, 'Narrative of Military Railways', 5.
28 'Notes on Embodiment of Burma Railways Battalion by Staff Captain', 12 January 1942, File 4, BFP; Micklem, *Transportation*, 199.
29 Foucar, 'Draft Narrative', 199, 200; Enclosure 29 Appendix: Burma Transportation Directorate Report, 2, WO/106/2654, PRO; R. P. Pakenham-Walsh, *History of the Corps of Royal Engineers*, vol. IX (Chatham: The Institution of Royal Engineers, 1958), 165; Kirby, *The War Against Japan*, vol. 2, 466; Reginald Melville Ward Lowe,

'Some Memories of Burma and the Burma Railways'. Anglo-Burmese Library, http:// www.ablmembersarea.com/reginald-melville-ward-lowe.html [accessed 26 April 2015].

30 Lt. Col. Seymour Williams, 'Report on the Work of the No. 2 Docks Groups I.E. in Rangoon during the Period 12-1-42 to 7-3-42', 1, File 5, BFP.

31 Biddulph, 'Narrative of Military Railways', 3.

32 Williams, 'Report on the Work of the No. 2 Docks Groups IE', 1.

33 Ibid.

34 Biddulph, 'Narrative of Military Railways', 4.

35 Ibid., Biddulph to Goddard, 27 June 1942, 2, File 4, BFP.

36 McKenna, *The Railway Workers 1840–1970*, 63–4.

37 IWM Milne Papers, Milne to Foucar, 23 March 1943, 1; Milne, 'The Storey of Fall of Rangoon and Departure Therefrom', 3–4, Milne Papers, IWM.

38 Biddulph, 'Narrative of Military Railways', 4.

39 Enclosure 29 Appendix: Burma Transportation Directorate Report, 2, WO/106/2654, PRO; Foucar, 'Draft Narrative', 199–200; Pakenham-Walsh, *History of the Corps of Royal Engineers*, vol. 13, 165.

40 Biddulph, 'Narrative of Military Railways', 4.

41 Kirby, *The War Against Japan*, vol. 2, 467.

42 Biddulph, 'Narrative of Military Railways', 4–5.

43 Ibid., 6.

44 Ibid., Chance to Biddulph, 7 February 1942, File 4, BFP.

45 Ibid.; Biddulph, 'Narrative of Military Railways', 6.

46 Kirby, *The War Against Japan*, vol. 2, 467.

47 Edward Warren Caulfield Sandes, *Indian Engineers, 1939–1947* (Kirkee: Institution of Military Engineers, 1956), 212.

48 Biddulph, 'Report on the Burma Railways in the Context of War', 6, 35.

49 Walter Hingston, *Never Give Up. Being Volume 5 of The History of the King's Own Yorkshire Light Infantry 1919–1942* (London: Lund Humphries, 1950), 176; Biddulph, 'Report on the Burma Railways', 5, 10–11.

50 Biddulph, 'Report on the Burma Railways', 19, 35.

51 James Lunt, *The Retreat from Burma, 1941–1942* (Trowbridge: David & Charles, 1989), 64.

52 Hingston, *Never Give Up*, 176.

53 Ibid., 155; James Noble MacKay, *History of the 7th Duke of Edinburgh's Own Gurkha Rifles* (Edinburgh: William Blackwood & Sons, 1962), 69.

54 Biddulph, 'Narrative of Military Railways', 3.

55 D. G. Lacy Scott, 'Burma: The Great Withdrawal', *Burma Campaign Society Newsletter* 15 (March, 2010), 13.

56 Confidential Report on A. Fitzherbert, L/SG/11/8/89, IOR.

57 Dubashi's account is quoted at length in Sandes, *Indian Engineers, 1939–1947*, 215–16.

58 2nd Infantry Brigade, Operations Instruction No. 4, 14 February 1942, 2, WO172/548, PRO.

59 Jennings, [Conditions in the Thaton District], 2; Hingston, *Never Give Up*, 182; Biddulph, 'Report on the Burma Railways, 19; Sandes, *Indian Engineers, 1939–1947*, 216–17, 219.

60 Lady Dorman-Smith, the governor's wife, described the news as 'grim'. Lady Dorman Smith Diary, 1941–2, entry for 16 February 1942, Mss Eur/215E/41, IOR.

61 Mole, *The Temple Bells Are Calling*, 129; G. M. O. Davy, *The Seventh and Three Enemies: The Story of World War II and the 7th Queen's Own Hussars* (Cambridge: W. Heffer & Sons, 1952), 210.

62 See the enclosed, hand-drawn maps in Burma, Note on Communications from Moulmein to Rangoon, WO106/2640, PRO.

63 Edward Warren Caulfield Sandes, *From Pyramid to Pagoda: The Story of the West Yorkshire Regiment (The Prince of Wales's Own) in the War, 1939–45 and Afterwards*. With a foreword by Field-Marshal Sir William Slim (London: F. J. Parons, 1952), 21.

64 See the enclosed, hand-drawn maps in Burma, Note on Communications from Moulmein to Rangoon, WO106/2640, PRO.

65 Ibid.

66 Biddulph, 'Report on the Burma Railways', 40.

67 Sandes, *From Pryamid to Pagoda*, 21.

68 Wavell to Hartley, 23 February 1942, Hutton 2–13, Hutton Papers.

69 Ibid.

70 Ibid.; Wavell to Hartley, 23 February 1942, Hutton 2–13, Hutton Papers; Sandes, *From Pryamid to Pagoda*, 22.

71 Biddulph, 'Narrative of Military Railways', 21.

72 Hutton to ABDACOM, 18 February 1942, Hutton 2–13, Hutton Papers.

73 Soden, 'Report by the Director of Movement Control', 5; Foucar, 'Draft Narrative', 90; Johnson, 'A Review of the Burma Railways', 179; 'K', 'Burma in My Life Time', 15; Fitzpatrick, *No Mandalay No Maymyo (79 Survive)*, 80.

74 Micklem, *Transportation*, 203.

75 Foucar, 'Draft Narrative', 90.

76 I have found especially useful for this discussion Map 10, 'Pegu Road Block 7 March 1942', included in James Noble MacKay, *A History of the 4th Prince of Wales's Own Gurkha Rifles* (Edinburgh: William Blackwood & Sons, 1952), vol. 3, 92.

4 Militarization and the Rangoon to Prome Branch

1 Biddulph, 'Narrative of Military Railways', 4.

2 Manton to Biddulph, 19 March 1942, File 4, BFP.
3 War Diary of the No. 3 RCMG, for January to July 1942, 2, 5; Foucar, 'Draft Narrative', 200. W. H. Prendergast, 'A Galway Engineer in Assam', http://places.galwaylibrary.ie/history/chapter70.html (accessed 28 November 2014).
4 William Hillary Prendergast obituary, 232.
5 War Diary of the No. 3 RCMG, for January to July 1942, 2.
6 Ibid., 2–3, 5; WH Prendergast, 'Railway Man in Burma', 2, WO/172/1021/3, PRO; whether this unit refers to Lieutenant Carre's group, arriving in early March, remains unclear.
7 War Diary of the No. 3 RCMG, for January to July 1942, 6.
8 Biddulph to Manton, 1 March, 1942, 2, File 4, BFP.
9 Governor Dorman-Smith, 'Comments on General Hutton's Despatches'.
10 Biddulph, 'Narrative of Military Railways', 6.
11 Biddulph, 'Memorandum on Military Transportation Organisation', 13 February 1941, File 4, BFP.
12 Biddulph to Various Heads of Agencies and Departments, 14 February, 1, File 4, BFP.
13 Ibid.
14 Biddulph, 'Narrative of Military Railways', 6.
15 Dorman-Smith, *Report on the Burma Campaign, 1941–1942*, 171.
16 Prendergast, 'Railway Man in Burma', 1.
17 Quoted in Foucar, 'Draft Narrative', 200.
18 Johnson, 'A Review of the Burma Railways', 179.
19 Biddulph, 'Narrative of Military Railways', 8; Kirby, *The War Against Japan*, vol. 2, 468; Thompson, *The Official History of New Zealand in World War II*, 277.
20 Burma Army to ABDACOM, 24 February 1942, Hutton 2–3, Hutton Papers.
21 Goddard, 'Army Administration', 29 March 1942, File 2–39, 1, 4, Hutton Papers.
22 Dorman-Smith, *Report on the Burma Campaign, 1941–1942*, 172; Pakenham-Walsh, *History of the Corps of Royal Engineers*, Vol. IX, 165; Prendergast, 'Railway Man in Burma', 1; Kirby, *The War Against Japan*, vol. 2, 468; Milne, 'The Storey of Fall of Rangoon and Departure Therefrom', 3; War Diary of the No. 3 RCMG, for January to July 1942, 2; Foucar, 'Draft Narrative', 200; Biddulph, 'Narrative of Military Railways', 8.
23 Biddulph to Manton, 1 March 1942, 1, File 4, BFP.
24 Foucar, 'Draft Narrative', 200; Biddulph, 'Narrative of Military Railways', 8; see Appendix B to Sir John Rowland, List of Railway Dates and Documents, L/WS/1/1512, 3, IOR, BL, London, UK.
25 Biddulph, 'Burmah Notebook', 9, File 5, BFP.
26 Dorman-Smith, *Report on the Burma Campaign, 1941–1942*, 172; Statement by E. I. Milne for T. L. Hughes, 5 September 1942, 1; Mains, *Soldier with Railways*,

57; *Supplement to the London Gazette*, 28 October 1942, 4687; Milne, 'The Storey of Fall of Rangoon and Departure Therefrom', 3–4; *Railway Gazette* 77 (31 July 1942), 112; Recommendation for Hector Soord, WO/373/30, PRO; Confidential Report, A. F. Fitzherbert Personnel File, L/SG/11/8/89, IOR; Confidential Report on S. C. Bryant, 16 March 1944, L/SG/11/8/37, IOR; Confidential Report on C. C. R. Edwards, 29 February 1944, L/SG/11/8/78, IOR; Confidential Report on C. N. Blakeney, 22 October 1942, L/SG/11/8/27, IOR.

27 Milne to Foucar, 23 March 1943, 2; Dorman-Smith, 'Comments on General Hutton's Despatches'; Statement by E. I. Milne for T. L. Hughes, 5 September 1942, 2; Soden, 'Tour Notes', 28 February 1942, 1, WO/172/372, PRO.

28 F. H. Yarnold to Burma Office, 29 June 1943, 2, M/3/1597, Indian Office Records, BL; Biddulph, 'Narrative of Military Railways', 8; Dorman-Smith, *Report on the Burma Campaign, 1941–1942*, 172; Statement by E. I. Milne for T. L. Hughes, 5 September 1942, 1; Kirby, *The War Against Japan*, vol. 2, 468; War Diary of the No. 3 RCMG, for January to July 1942, 3; Prendergast, 'Railway Man in Burma', 2; Biddulph, 'Narrative of Military Railways', 8.

29 Johnson, 'A Review of the Burma Railways', 179; Statement by E. I. Milne for T. L. Hughes, 5 September 1942, 1; Milne, 'The Storey of Fall of Rangoon and Departure Therefrom', 3; Kirby, *The War Against Japan*, vol. 2, 468; *Hnit-taya-byi Mi-yahta 1877–1977*, 196; Cardew, 'Extracts from the Diary of Jack Cardew', 4.

30 Edward Law-yone, extracts from unpublished memoir, 2.

31 Biddulph, 'Report on the Burma Railways', 8.

32 H. G. Wilkie, *Report on the Effect of Denial and Demolition Schemes in Burma*, 35, WO/203/5707, PRO.

33 Recommendation for Hector Soord.

34 Johnson, 'A Review of the Burma Railways', 179; *Hnit-taya-byi Mi-yahta 1877–1977*, 196; Cardew, 'Extracts from the Diary of Jack Cardew', 1.

35 Biddulph, 'Report on the Burma Railways', 9; Johnson, 'A Review of the Burma Railways', 179, 206; Recommendation for Hector Soord; Wilkie, *Report on the Effect of Denial and Demolition Schemes in Burma*, 35.

36 Sandes, *Indian Engineers, 1939–1947*, 217.

37 Recommendation for award for Lt. Raymond Taylor Carr, WO/373/79, PRO; Pakenham-Walsh, *History of the Corps of Royal Engineers*, vol. 13, 165; Kirby, *The War Against Japan*, vol. 2, 468; Manton to Biddulph, 19 March 1942, File 4, BFP.

38 War Diary of the No. 3 RCMG, for January to July 1942, 3; WH Prendergast, 'Railway Man in Burma', 2; Biddulph to Manton, 1 March, 1942, 2, File 4, BFP.

39 Lunt, *The Retreat from Burma*, 170.

40 War Diary of the No. 3 RCMG, for January to July 1942, 3; Prendergast, 'Railway Man in Burma', 2.

41 G. T. Porter, 'The Lines behind the Lines in Burma', *Railway Magazine* 89, no. 548 (November & December 1943): 325–6.

42 Abraham, 'Details of a Tour in Burma in March 1942', 2, Abraham Papers, LHCMA.

43 'K', 'Burma in My Life Time', 16; Kirby, *The War Against Japan*, vol. 2, 468.

44 The 3 RCMG, War Diary for January to July 1942, 3; Foucar, 'Draft Narrative', 92; Alfred Wagg. *A Million Died! A Story of War in the Far East* (London: Nicholson & Watson, 1943), 53; John L. Christian, 'The Japanese Campaign for Burma December, 1941 – May, 1942', (1944), 12, WO/203/4651, PRO; Harold Alexander, *The Alexander Memoirs 1940–1945*. Edited by John North (London: Cassell, 1962), 92, 100; Brewitt, 'Military Railway Operation in the Defence of Rangoon, 25th February–8th March, 1942' 3, File 5, BFP.

45 PRO/WO/172/447 HQ 1st Burma Division War Diary, 2.

46 Biddulph, 'Narrative of Military Railways', 14.

47 Brewitt, 'Military Railway Operation in the Defence of Rangoon, 25th February-8th March, 1942' 3, File 5, BFP; Biddulph, 'Narrative of Military Railways', 8.

48 Alexander, *The Alexander Memoirs*, 95, 100.

49 Ronald Lewin, *Slim the Standardbearer: A Biography of Field-Marshal the Viscount Slim* (London: Wordsworth Editions, 1976), 82f.

50 Kirby, *The War Against Japan*, vol. 2, 466; Wilkie, *Report on the Effect of Denial and Demolition Schemes in Burma*, 25.

51 HQ, American Army Forces China, Burma and India, 'The Campaign in Burma (from March 1st to June 1st)', 1, WO/203/4651, PRO.

52 Andrus, *Burmese Economic Life*, 235; War Diary of the No. 3 RCMG, for January to July 1942, 5.

53 War Diary of the No. 3 RCMG, for January to July 1942, 5.

54 'Narrative Account of 1st Hy AA Regiment, BAF', 2.

55 Dorman-Smith, 'Comments on General Hutton's Despatches; 'Dorman-Smith, *Report on the Burma Campaign, 1941–1942*, 172; Rowland, List of Railway Dates and Documents, 2; Rowland, List of Railway Dates and Documents, 2; GOB, *Compendium of Governor's Acts Published in India – 1942–1945* (Rangoon: Superintendent, Government Printing, 1946), 8.

56 Dorman-Smith telegram to Secretary of State, dated 15 March 1942, Photo Eur/011, British Library, European Manuscripts Collection, London; Soden, 'Report by the Director of Movement Control', 9; Foucar, 'Draft Narrative', 200; Johnson, 'A Review of the Burma Railways', 206; *Railway Gazette International* 77 (1943), 586; Milne, 'The Storey of Fall of Rangoon and Departure Therefrom', 4; Mains, *Soldier with Railways*, 65. See Kirby, *The War Against Japan*, vol. 2, 468.

57 Defence Department, GOB to Burma Office, 29 June 1943, 1, M/3/1597, Indian Office Records, BL.

58 Ibid.

59 Rowland, List of Railway Dates and Documents, 2.

60 Law-yone, extracts from unpublished memoir, 2; Cardew, 'Extracts from the Diary of Jack Cardew', 6; Biddulph, 'Burmah Notebook', 9.

61 Statement by E. I. Milne for T. L. Hughes, 5 September 1942, 1; Milne, 'The Storey of Fall of Rangoon and Departure Therefrom', 4; GOB, *Annual Report of the Burma Railway Board 1945–46*, 1.

62 Confidential Report on G. H. Binnie, 10 March 1943, L/SG/11/8/25, IOR; Cardew, 'Extracts from the Diary of Jack Cardew', 7; War Diary of the No. 3 RCMG, for January to July 1942, 5–6; 'Report on R. D. Cant', 18 July 1947, LSG/11/8/44, IOR.

63 HQ 1st Burma Division War Diary, 2, WO/172/447, PRO; WH Prendergast, 'Railway Work during the Burma Campaign, 1942', WO/172/1021, 7, PRO; Edward Law-yone, *Golden Parasol: A Daughter's Memoir of Burma* (London: Vintage, 2014), 65.

64 Newton, 'Exodus from Burma', 23–4, Newton Papers, IWM.

65 Prendergast, 'Railway Work', 6.

66 C. M. Enriquez, *Burma Invaded 1942* (n.p.: Margaret P. Stanford, 2012), 138.

67 Quoted in Laura Tyson Li, *Madame Chiang Kai-shek: China's Eternal First Lady* (New York: Grove Press, 2007), 176.

68 Biddulph to Manton, 1 March 1942, 1, File 4, BFP.

69 John Maxwell Hamilton, *Journalism's Roving Eye: A History of American Foreign Reporting* (Baton Rouge: Louisiana State University Press, 2009), 313.

5 Kings of the road: Decentralization and local initiative on the railway

1 Atsuko Naono, *State of Vaccination: The Fight against Smallpox in Colonial Burma* (Himayatnagar: Orient BlackSwan, 2009).

2 Jonathan Saha, *Law, Disorder and the Colonial State: Corruption in Burma, c. 1900* (Houndmills: Palgrave Macmillan, 2013).

3 Keith G. Stewart, 'The Evolution of Command Approach' (Toronto: Defence Research and Development Canada, n.d.), 1–3; Stephen Bungay, 'The Road to Mission Command: The Genesis of a Command Philosophy', *British Army Review* 137 (2005), 22–9; H. P. Willmott, *When Men Lost Faith in Reason: Reflections on War and Society in the Twentieth Century* (London: Greenwood, 2002), 209; G. D. Sheffield, 'The Australians at Pozieres: Command and Control on the Somme, 1916', in *British General Staff: Reform and Innovation*, ed. David French & Brian Holden Reid (London: Routledge, 2004), 98; Christopher Pugsley, 'We Have Been Here Before: The Evolution of the Doctrine of Decentralised Command in the British Army, 1905–1989', *Sandhurst Occasional Papers* 9 (2011), 6–7.

4 Patrick Rose, 'Allies at War: British and US Army Command Culture in the Italian Campaign, 1943–1944', *Journal of Strategic Studies* 36, no. 1 (2013), 43.

5 Coward to Brewitt, 13 July 1942, 2–3.

6 HQ 1st Burma Division War Diary, 2, WO/172/447, PRO; Coward to Brewitt, 13 July 1942, 2.

7 Johnson, 'A Review of the Burma Railways', 179; Statement by E. I. Milne for T. L. Hughes, 5 September 1942, 1; Milne, 'The Storey of Fall of Rangoon and Departure Therefrom', 3; Kirby, *The War Against Japan*, vol. 2, 468; *Hnit-taya-byi Mi-yahta 1877–1977*, 196; Cardew, 'Extracts from the Diary of Jack Cardew', 4; Biddulph, 'Narrative of Military Railways', 9.

8 Lewis, 'Report at Toungoo', 19 March 1942, 1, File 5, BFP; Biddulph, 'Narrative of Military Railways', 9; Coward to Brewitt, 13 July 1942, 3.

9 Coward to Brewitt, 13 July 1942, 1.

10 Lewis, 'Report at Toungoo', 1, 3; Biddulph, 'Narrative of Military Railways', 9.

11 Biddulph, 'Narrative of Military Railways', 9.

12 Lewis, 'Report at Toungoo', 2.

13 Ibid., 1–2.

14 Sandes, *From Pyramid to Pagoda*, 22; HQ 1st Burma Division War Diary, 2, WO/172/447, PRO; Wagg, *A Million Died*, 55–6; *Railway Gazette* 77 (31 July 1942), 112.

15 Confidential Report on Mr J. D. Lewis, 10 August 1943, L/SG/11/8/148, IOR.

16 Coward to Brewitt, 13 July 1942, 6.

17 H. P. Munro, 'Retreat December 1941 to May 1942: 24 (RB) Field Company and Malerkotla Field Company'. In *The Royal Bombay Sappers & Miners, 1939–1947* (Bombay: Royal Bombay Sappers & Miners Officers' Association, 1999), 276; R. Hudson (comp.), 'Diary of the Malerkotla Field Company, Indian State Forces, October 1941–May 1942', 11, File No. 13, REM World War II Unit War Diaries, Royal Engineers Museum Archive, Gillingham, UK.

18 Ibid., 6–7.

19 HQ 1st Burma Division War Diary, 2; Wagg, *A Million Died*, 55–6; *Railway Gazette* 77 (31 July 1942), 112; Sandes, *Indian Engineers, 1939–1947*, 226.

20 Munro, 'Retreat December 1941 to May 1942', 276; Hudson (comp.), 'Diary of the Malerkotla Field Company, Indian State Forces, October 1941–May 1942', 11; Wagg, *A Million Died*, 55–56; *Railway Gazette* 77 (31 July 1942), 112. Zyoichi Saito, 'Diary of a Company Commander', in *Railwaymen in the War: Tales by Japanese Railway Soldiers in Burma and Thailand 1941–47*, ed. and tr. Kazuo Tamayama (London: Palgrave Macmillan, 2005), 69.

21 Wagg, *A Million Died*, 56; Lewis, 'Report at Toungoo', 19 March 1942, 3, File 5, BFP; *Supplement to the London Gazette*, 29 January 1943, 554.

22 Lewis, 'Report at Toungoo', 19 March 1942, 3.

23 Ibid., 3–4; Carr to Prendergast, 21 March 1942, 1–3, BFP; Biddulph, 'Narrative of Military Railways', 14, File 4, BFP.

24 Coward to Brewitt, 13 July 1942, 8, 14; Award Citation for Lt. Raymond Taylor Carr, WO/373/79, PRO; Carr to Prendergast, 21 March 1942, 2–3, BFP; Lewis, 'Report at Toungoo', 19 March 1942, 3, File 4, BFP.

25 Carr to Prendergast, 21 March 1942, 1–4.
26 Foucar, 'Draft Narrative', 202; Carr to Prendergast, 21 March 1942, 3, BFP; Enriquez, *Burma Invaded 1942*, 125.
27 Newton, 'Exodus from Burma', 23–4.
28 Coward to Brewitt, 13 July 1942, 1.
29 Biddulph Instructions to Brown, 27 March 1942, 1, File 5, BFP.
30 Coward to Brewitt, 13 July 1942, 2–3.
31 Enriquez, *Burma Invaded 1942*, 108; John E. Ausland [under pseudonym, 'The Old Gray Major'], 'Railway South of the Clouds'. *Ex-CBI Roundup* (November 1955), 15; Enriquez, *Burma Invaded 1942*, 124, 126.
32 Ibid., 114, 175.
33 Munro, 'Retreat December 1941 to May 1942', 276; Saito, 'Account of Lieutenant Zyoichi Saito, Commander of 4th Company', 70.
34 Carr comments in Lewis, 'Report at Toungoo', 19 March 1942, 4–5, File 5, BFP.
35 Pearn, 'Report on the Civil Evacuation', 34; Munro, 'Retreat December 1941 to May 1942', 276; Saito, 'Account of Lieutenant Zyoichi Saito, Commander of 4th Company', 70; C. P. Brewitt, 'The Chinese Army in Burma', 1, WO/203/5721, PRO.
36 Letter to E. Foucar, 20 August 1943, 2, WO/203/5731, PRO; Enriquez, *Burma Invaded 1942*, 125; Wagg, *A Million Died*, 79.
37 Ibid., 114.
38 Brewitt, 'The Chinese Army in Burma', 1; Soden, 'Report by the Director of Movement Control', 8; Foucar, 'Draft Narrative', 114.
39 Confidential Report on Mr J. D. Lewis, 10 August 1943, L/SG/11/8/148, IOR; Coward to Brewitt, 13 July 1942, 4.
40 Pearn, 'Report on the Civil Evacuation', 34; Coward to Brewitt, 13 July 1942, 5.
41 War Diary – CLM, 5, WO/203/5721, PRO.
42 Biddulph, 'Narrative of Military Railways', 18.
43 Soden, 'Report by the Director of Movement Control', 5; Enriquez, *Burma Invaded 1942*, 56; Dorman-Smith, *Report on the Burma Campaign, 1941–1942*, 171; Sein Tin, *Sit-atwìn Neí-sin Hmat-tàn* (Rangoon: Pugan Sa-oub, 1966), 274.
44 Biddulph, 'Report on the Burma Railways', 7–8.
45 Soden, 'Report by the Director of Movement Control', 5; Johnson, 'A Review of the Burma Railways', 206; Dorman-Smith, *Report on the Burma Campaign, 1941–1942*, 170–1; Cardew, 'Extracts from the Diary of Jack Cardew', 6.
46 War Diary – CLM, 5, 7, 8, WO/203/5721, PRO; Coward to Brewitt, 13 July 1942, 3–5, 7.
47 Coward to Brewitt, 13 July 1942, 3, 5.
48 Ibid., 5; War Diary – CLM, 5, WO/203/5721, PRO.
49 Coward to Brewitt, 13 July 1942, 15.
50 Ibid., 6, 8; War Diary – CLM, 5, 7, 8, WO/203/5721, PRO.

51 Coward to Brewitt, 13 July 1942, 8.

52 Ibid., 8, 14; Award Citation for Lt. Raymond Taylor Carr, WO/373/79, PRO.

53 War Diary – CLM, 6, 8; Coward to Brewitt, 13 July 1942, 3–4, 6.

54 Coward to Brewitt, 13 July 1942, 8–9; War Diary – CLM, 8, WO/203/5721, PRO.

55 Ibid.; Brewitt, 'The Chinese Army in Burma', 2; 'Note on Working Thazi Loco Shed', 2.

56 'Note on Working Thazi Loco Shed', 2.

57 Brewitt, 'The Chinese Army in Burma', 2.

58 'Note on Working Thazi Loco Shed', 1–2.

59 Milne to Foucar, 23 March 1943, 3.

60 Ibid., 4.

61 J. V. Davidson-Houston, 'Some Personal Experiences in Burma, 1942', *Journal of the Royal United Service Institution* 88, no. 550 (1943), 124.

62 Foucar, 'Draft Narrative', 114; Graham Dunlop, *Military Economics, Culture and Logistics in the Burma Campaign, 1942–1945* (London: Pickering & Chatto, 2009), 23.

63 Coward to Brewitt, 13 July 1942, 10.

64 'Notes on Report Received from Mr Waters, Civil Transport Officer', circa 16 April 1942, 1, File 5, BFP; Coward to Brewitt, 13 July 1942, 11–12.

65 Coward to Brewitt, 13 July 1942, 12; 'Notes on Report Received from Mr Waters', 1–2.

66 'Notes on Report Received from Mr Waters', 1.

67 Ibid.; Coward to Brewitt, 13 July 1942, 12–13.

68 Burma, *Annual Report of the Burma Railway Board, 1940–41*, 29.

69 'Report on an Enquiry into Railway Accidents in Burma', 9 September 1942, M/3/541, IOR.

70 Burma, *Annual Report of the Burma Railway Board, 1940–41*, 29.

71 Coward to Brewitt, 13 July 1942, 8.

72 Ibid., 15.

73 Ibid.

74 Stowe, *They Shall Not Sleep*, 96–7.

75 Burma Corps G Branch War Diary for April 1942, 5, WO/172/403, PRO.

76 Jack Belden, *Retreat with Stilwell* (New York: Da Capo Press, 1975), 174.

77 Even in better circumstances, when the Prussians were winning the 1870 Franco-Prussian War there was a tendency for the Prussians to blame any accident on the French railways they controlled on sabotage. Van Creveld, *Supplying War: Logistics from Wallenstein to Patton*, 96.

78 This is implied at paragraph 153 in Burma, *Annual Report of the Burma Railway Board, 1940–41*, 32–3.

79 Mains, *Soldier with Railways*, 59.

80 Porter, 'The Lines Behind the Lines in Burma', 326.
81 Prendergast, 'Railway Work', 2; 'The Burma Railways during Invasion', 216; Foucar, 'Draft Narrative', 201; Davidson-Houston, 'Some Personal Experiences in Burma, 1942', 125.
82 Hilda R. Corpe, *Prisoner beyond the Chindwin* (London: Arthur Barker, 1955), 41–2; Oung, *A Twentieth Century Burmese Matriarch*, 95.
83 William Slim, *Defeat into Victory* (London: Cassell, 1956), vol. 1, 30.
84 'Note on Working Thazi Loco Shed', 2.
85 Ibid.
86 'Notes on Report Received from Mr Waters', 1.
87 Ibid., 2.
88 Ibid.
89 Ibid., 1–2.
90 Ibid., 13.
91 Coward to Brewitt, 13 July 1942, 13–14.
92 'Note on Working Thazi Loco Shed', 4–6.
93 Ibid., 6.
94 Coward to Brewitt, 13 July 1942, 14.
95 Raina (ed.), *Campaigns in the Eastern Theatre*, 157; Foucar, 'Draft Narrative', 147.
96 Ausland, 'Railway South of the Clouds', 15; Paul LeRoy Jones, 'The Withdrawal from Burma and the Stillwell Walkout', *Ex-CBI Roundup* (May, 1992), Web Version, http://CBI-theater-3.home.comcast.net [accessed 1 August 2014]; Belden, *Retreat with Stilwell*, 174.
97 Belden, *Retreat with Stilwell*, 211–12, 215; Jones, 'The Withdrawal from Burma and the Stillwell Walkout'.
98 David Donnisson, *Last of the Guardians: A Story of Burma, Britain and a Family*, (Newtown: Cyhoeddwyr y Superscript, 2005), 248–9.
99 Slim, *Defeat into Victory*, 82.

6 The technical limits of the railway: The Mandalay to Lashio Branch

1 Goddard, 'Army Administration', 29 March 1942, File 2–39, 2, Hutton Papers.
2 Foucar, 'Draft Narrative', 201; Mains, *A Soldier with Railways*, 63; Fred Millem, 'Rangoon Battalion', introduced by Patricia Pringle. http://www.bbc.co.uk/history/ww2peopleswar/categories/c55521/ [accessed 9 September 2017].
3 Biddulph, 'Narrative of Military Railways', 19.
4 'Note on working Ywataung as a Locomotive Depot', 1–2, File 5, BFP.
5 Ibid., 2–4, File 5.

6 Ibid., 5–6.
7 Enriquez, *Burma Invaded 1942*, 108; Cardew, 'Extracts from the Diary of Jack Cardew', 5.
8 Cardew, 'Extracts from the Diary of Jack Cardew', 10, 12.
9 Ibid., 10.
10 Biddulph to Goddard, 27 June 1942, 1–2, File 4, BFP.
11 Ibid.
12 Dorman-Smith, *Report on the Burma Campaign, 1941–1942*, 172; Dorman-Smith, 'Comments on General Hutton's Despatches'.
13 Extract from a Letter Written from Hotel Cecil, Simla, India, 6 September 1942, 1, Mss Eur/F/481/3, IOR.
14 John Wise to T. Hutton, Simla, 28 February 1944, Hutton 3–9, Hutton Papers.
15 GOB, *Annual Report of the Burma Railway Board 1945–46*, 1; Dorman-Smith, *Report on the Burma Campaign, 1941–1942*, 172; Statement by E. I. Milne for T. L. Hughes, 5 September 1942, 1.
16 Extract from a Letter Written from Hotel Cecil, Simla, India, 6 September 1942, 1.
17 Cardew, 'Extracts from the Diary of Jack Cardew', 11.
18 Ibid.
19 Letter to E. Foucar, 23 July 1943, 4.
20 Appendix F to Sir John Rowland, List of Railway Dates and Documents, L/WS/1/1512, 4, IOR, BL, London, UK.
21 Ibid.
22 Extract from a Letter Written 13 August 1942, 1.
23 Pearn, 'Report on the Civil Evacuation', 39, 44; Dorman-Smith, *Report on the Burma Campaign, 1941–1942*, 172; Porter, 'The Lines behind the Lines in Burma', 326; Millem, 'Rangoon Battalion'.
24 J. S. Vorley, 'Report on the Work of the Department of Civil Evacuation and Welfare', M/3/955, IOR; the reference to free travel on the Burma Railways was made by Christopher Lorimer in 'East India Saga: A Family Story 1732–1944', 436, Cambridge University South Asian Archive.
25 Extract from a Letter Written 13 August 1942, 1.
26 These observations are made by Enriquez, *Burma Invaded 1942*, 204–5.
27 Wilfred Burchett, *Bombs over Burma* (London: F. W. Cheshire, 1944), 102–3.
28 Pearn, 'Report on the Civil Evacuation', 39, 44; Maurice Maybury, *Heaven-Born in Burma*, vol. 2, *Flight of the Heaven-Born* (Somerset: Folio Hadspen, 1985), 154–6.
29 Charles Haswell Campagnac, *The Autobiography of a Wanderer in England & Burma* (Raleigh, NC: Lulu Enterprises, 2011), 286.
30 Belden, *Retreat with Stilwell*, 202–3.
31 Ibid.
32 Ibid., 2.

33 Wilkie, *Report on the Effect of Denial and Demolition Schemes in Burma*, 36.
34 Dorman-Smith, *Report on the Burma Campaign, 1941–1942*, 173.
35 Law-yone, extracts from unpublished memoir, 5.
36 Foucar, 'Draft Narrative', 147.
37 Dorman-Smith, *Report on the Burma Campaign, 1941–1942*, 173.
38 Foucar, 'Draft Narrative', 147; Soden, 'Report by the Director of Movement Control', 10.
39 Extract from a Letter Written from Hotel Cecil, Simla, India, 6 September 1942, 1, MSS Eur/F/481/3, IOR; Extract from a Letter Written 13 August 1942, 1, MSS Eur/F/481/2, IOR; Statement by E. I. Milne for T. L. Hughes, 5 September 1942, 1; Milne, 'The Storey of Fall of Rangoon and Departure Therefrom', 1; Biddulph, personal notepad, 108, BFP.
40 Johnson actually says 100, but from the official demolitions report it is clear that he was rounding numbers up. Johnson, 'A Review of the Burma Railways', 206; Wilkie, *Report on the Effect of Denial and Demolition Schemes in Burma*, 35.
41 Enriquez, *Burma Invaded 1942*, 208.
42 Lowe, 'Some Memories'; Milne, 'The Storey of Fall of Rangoon and Departure Therefrom', 1; Extract from a Letter Written from Hotel Cecil, Simla, India, 6 September 1942, 1, MSS Eur/F/481/3, IOR; Extract from a Letter Written 13 August 1942, 1, MSS Eur/F/481/2, IOR; Statement by E. I. Milne for T. L. Hughes, 5 September 1942, 1; Milne, 'The Storey of Fall of Rangoon and Departure Therefrom', 1; Biddulph, personal notepad, 108, BFP.
43 Wilkie, *Report on the Effect of Denial and Demolition Schemes in Burma*, 35; Johnson, 'A Review of the Burma Railways', 206; idem, 'Report on Burma Railway Demolitions', 2; Pearn, 'Report on the Civil Evacuation', 39; Cardew, 'Extracts from the Diary of Jack Cardew', 12–13.
44 Dorman-Smith, *Report on the Burma Campaign, 1941–1942*, 173.
45 Johnson, 'A Review of the Burma Railways', 206; Foucar, 'Draft Narrative', 155, 174; Coward to Brewitt, 13 July 1942, 14.

7 Dark territory: The Myitkyina Branch, late April to early May 1942

1 Davy, *The Seventh and Three Enemies*, 284.
2 Kirby, *The War Against Japan*, Vol. 2, 210; Davy, *The Seventh and Three Enemies*, 291.
3 'Report on the Circumstances under which Major Eecles and Captain Gimmingham, 56 Field Company, were Captured by the Enemy', enclosure to Burma Divisional Sappers and Miners (27 February 1942 to 31 May 1942), File No. 13, REM World War II Unit War Diaries, Royal Engineers Museum Archive,

Gillingham, UK; 'Account of the Crossing of the Irrawaddy River by 1 Burma Division', enclosure to Burma Divisional Sappers and Miners (27 February 1942 to 31 May 1942), 2–3, File No. 13, REM World War II Unit War Diaries, Royal Engineers Museum Archive, Gillingham, UK; Hudson (comp.), 'Diary of the Malerkotla Field Company, Indian State Forces, October 1941-May 1942', 24; 'Account of the Crossing of the Irrawaddy River by 1 Burma Division', 2–3; Alan Warren, *Burma 1942: The Road from Rangoon to Mandalay* (London: Continuum, 2011), 222–3; Mole, *The Temple Bells Are Calling*, 143–5; Davy, *The Seventh and Three Enemies*, 292.

4 Letter to E. Foucar, 23 July 1943, 7.

5 Donnisson, *Last of the Guardians*, 249.

6 Joan Robertson, *Maymyo More Far: A Walk out of Burma 1942* (Banbury: Norman Hudson, 1999), 63.

7 Foucar, 'Draft Narrative', 175.

8 Davidson-Houston, *Armed Pilgrimage*, 269; idem, 'Some Personal Experiences in Burma, 1942', 125; Lathrop, 'Chinese Nationalist Troops in Burma', 422f; Dr John Grindlay, quoted in Lathrop, 'Dateline: Burma', 33; Fred Eldridge, *Wrath in Burma: the Uncensored Story of General Stilwell and International Maneuvers in the Far East* (Garden City: Doubleday, 1946), 92; Hingston, *Never Give Up*, 228; HQ, American Army Forces China, Burma and India, 'The Campaign in Burma (from March 10th to June 1st)', 26–7; Stillwell Diary entry, 1 May 1942.

9 Col. C. G. Stewart, 'The Trek from Burma', 06/114/1, 7, Imperial War Museum, London.

10 Davidson-Houston, 'Some Personal Experiences in Burma, 1942', 124; Davidson-Houston, *Armed Pilgrimage*, 270–2.

11 Holmston to Biddulph, 29 April 1942, BFP.

12 Biddulph, 'Note on the Personnel to be Evacuated', 30 April 1942, in Biddulph, personal notepad, 126, BFP.

13 About 18 April, some of the patients of the Chinese ambulance train had drawn revolvers and gone into the engine box where they threatened to shoot the engineers if the engine stopped. The driver kept the engine going until it ran out of water and blew up, the engineers fleeing afterwards, leaving the Chinese to their fate. J. S. Vorley and H. M. Vorley, *The Road from Mandalay* (Hernes Keep: Wilton 65, 2002), 83.

14 Ibid., 83–4; Vorley, 'Report on the Work of the Department of Civil Evacuation and Welfare in Burma', M/3/955, 43, IOR, BL; Mitchell, 'Misconduct of Chinese Troops in Burma', 30.

15 Biddulph, 'Report on the Burma Railways', 29

16 Soden, 'Report by the Director of Movement Control', 10–11; Johnson, 'A Review of the Burma Railways', 206.

17 Dorman-Smith unpublished memoirs, Mss Eur/E215/32b, IOR.
18 Biddulph, 'Report on the Burma Railways', 30; Enriquez, *Burma Invaded 1942*, 229; Vorley and Vorley, *The Road from Mandalay*, 93.
19 Cardew, 'The Diary of Jack Cardew', 15–16; War Diary of the No. 3 RCMG, for January to July 1942, 6; Extract from a Letter Written from Hotel Cecil, Simla, India 6 September 1942, 1, Mss Eur/F/481/3, IOR; Extract from a Letter Written 13 August 1942, 1, Mss Eur/F/481/2, IOR; Statement by E. I. Milne for T. L. Hughes, 5 September 1942, 1; Burchett, *Bombs over Burma*, 102–3.
20 Johnson, 'A Review of the Burma Railways', 206; Dorman-Smith, *Report on the Burma Campaign, 1941–1942*, 173.
21 Alexander orders, 6 May 1942 and Unknown orders, 7 May 1942, both in Biddulph, personal notepad, 156, BFP; Indian Army to Biddulph, 9 May 1942, BFP; Diary of Captain W. A. Hutcheon, ABRO, attached Royal Marines on the River Patrol and Covering Movements from attack on Prome, 4 in WO/203/5691, PRO; Biddulph note, 10 May 1942, File 5, BFP.

8 After Burma: The militarization of railways in India

1 Biddulph, 'Memo on Transportation in India', 8–9.
2 Alexander, *The Alexander Memoirs 1940–1945*, 93.
3 Quoted in Adrian Cowells, 'Adventure on the Stilwell Road', *Popular Mechanics* 108, no. 5 (November, 1957), 83.
4 Dunlop, *Economics, Culture, and Logistics in the Burma Campaign*, 193–4.
5 Postscript to Abraham, 'Railway Operation in Eastern India', 1, Box 1, Abraham Papers, LHCMA.
6 Ibid.
7 Biddulph, 'Memo on Transportation in India', 3.
8 E. C. V. Foucar, 'The Draft Narrative', *Journal of the United Service Institution of India* 72, no. 308 (1945), 59–72; no. 309 (1945), 191–202.
9 Foucar, Memorandum 'Records and History of the first Burma Campaign', i–iii.
10 Foucar, 'Draft Narrative', 202; Dorman-Smith, *Report on the Burma Campaign, 1941–1942*, 172.
11 Ibid.
12 Information on the reasons for Manton's resignation and departure from India has been gratefully provided by Manton's grandson. Chris Manton, personal communication, 14 November 2014. Information on Manton's replacement by Gardiner is provided in Micklem, *Transportation*, 181. On Vickers's appointment, see *London Gazette* (20 March 1942), 1275.

13 Bayly & Harper, *Forgotten Armies: Britain's Asian Empire and the War with Japan*, 164.
14 HQ, American Army Forces China, Burma and India, 'The Campaign in Burma (from March 1st to June 1st)', 33.
15 Charles F. Romanus & Riley Sunderland, *United States Army in World War II China-Burma-India Theater: Stilwell's Mission to China* (Washington, DC: Center of Military History, 1987), 8, 46.
16 Ibid., 44–5.
17 Slim, *Defeat into Victory*.
18 Quoted in Biddulph, 'Military Transportation in India' 1.
19 Biddulph, 'Military Transportation in India', 1.
20 Biddulph to McMullen, 25 November 1942, 1, File 4, BFP.
21 Ibid.
22 Biddulph to McMullen, 25 November 1942, 1.
23 Biddulph, 'Narrative of Military Railways', 21.
24 Ibid., 22, File 4.
25 Ibid.
26 Biddulph, 'Notes on Memorandum on Movements in India', 3.
27 Ibid., 1–2.
28 Ibid., 1.
29 Biddulph, 'Narrative of Military Railways', 23.
30 Ibid.
31 Foucar, 'Draft Narrative', 202; Dorman-Smith, *Report on the Burma Campaign, 1941–1942*, 172.
32 Biddulph, 'Memo on Transportation in India', 4.
33 Confidential Report on Mr C. H. Whapham, L/SG/11/8/284, IOR.
34 Confidential Report on C. C. R. Edwards, 29 February 1944, L/SG/11/8/78, IOR; Report on C. C. R. Edwards, 17 July 1947, L/SG/11/8/78, IOR.
35 Report on Maj. T. Douglas Lee, 19 April 1944, L/SG/11/8/143, IOR.
36 Report on C. N. Blakeney, 7 October 1946, L/SG/11/8/37, IOR; Confidential Report on C. N. Blakeney, 16 March 1944, L/SG/11/8/37, IOR.
37 F. H. Yarnold to Burma Office, 29 June 1943, 2, M/3/1597, Indian Office Records, BL.
38 Ibid.
39 Lt. Col. JD Lewis, Draft letter to GOB, Telegram May 4 1945; Lewis Note, 30 May 1945; FWH Smith to Lewis, 22 August 1945, M/4/1673, IOR; Confidential Report on Mr J. D. Lewis, 10 August 1943, L/SG/11/8/148, IOR.
40 Ibid.
41 Biddulph, 'Memo on Transportation in India', 4.
42 Ibid.

43 Lowe, 'Some Memories'.
44 War Diary for the 172 RWC, PRO/WO/172/2962, National Archives, London; From O.C. No. 172 Ind. Railway W/S/ Coy to the 2nd Echelon, G.H.Q (i), Jhansi, 6 June 1942, PRO/WO/172/2962, National Archives, London; Verma & Anand, *The Corps of Indian Engineers, 1939–1947*, 254, 459.
45 SEAC HQ Memorandum, 5 June 1944, 'Transportation Personnel in Burma', WO 203/2867, 1, PRO; GHQ, India to SEAC HQ, 21 August 1944, regarding Transportation Personnel Ex-Burma, WO 203/2867, 1, PRO.
46 Micklem, *Transportation*, 183; Report on A. F. Fitzherbert, L/SG/11/8/89, IOR; 'Report on R. D. Cant'.
47 Report on A. F. Fitzherbert, L/SG/11/8/89, IOR; Report on C. N. Blakeney, 19 September 1946, L/SG/11/8/27, IOR; Report on G. H. Binnie, 19 July 1947, L/SG/11/8/25, IOR; Report on Mr C. H. Morgan, 30 August 1947, L/SG/11/8/178, 2, IOR; Report on E. J. R. Markwick, 19 July 1947, L/SG/11/8/166, IOR; Report on Mr N. McAlister, 19 July 1947, L/SG/11/8/156, 1, IOR.
48 JCB Wakeford, CRC, Burma, Report on Mr C. P. Brewitt, 19 July 1947, L/SG/11/8/34, IOR; Rear HQ, 11th Army Group Memorandum 'Transfer of Officers from GHQ (I) to Transportation Directorate, Burma', 24 September 1944, Appendix B, WO 203/2942, 1, PRO.
49 Prendergast, 'Railway Man in Burma'.
50 R. Gardiner, 'The Assam Lines of Communication', *Royal Engineers Journal* 62 (1948): 98.
51 Ibid., 101; Holden, *Let Smoke Steam*, 69.
52 Mains, *Field Security Very Ordinary Intelligence*, 63; Holden, *Let Smoke Steam*, 80.
53 Gardiner, 'The Assam Lines of Communication', 98; Mains, *Field Security Very Ordinary Intelligence*, 58–9.
54 Ibid., 55.
55 G. N. Russell & R. Gardiner, 'Planning, Development and Operation of the Assam Lines of Communication, 1941–1945', Part 2, 1–2, File 3, Durnford Papers, LHCMA, London, UK.
56 Ibid., Part 2, 3.
57 Gardiner, 'The Assam Lines of Communication', 98.
58 Russell & Gardiner, 'Planning, Development and Operation of the Assam Lines of Communication, 1941–1945', Part 2, 2–3.
59 Gardiner, 'The Assam Lines of Communication', 98.
60 Kirby, *The War Against Japan*, vol. 3, 459–60.
61 Mains, *Field Security Very Ordinary Intelligence*, 46–50; Holden, *Let Smoke Steam*, 46–7.
62 Mains, *Field Security Very Ordinary Intelligence*, 55–7; Holden, *Let Smoke Steam*, 74–5.

63 *Military Railways: Rule Book for Use by Indian Troops in India and Other Eastern Countries* (Moghalpura: North Western Railway Press, 1944), 32.
64 A. R. Godwin-Austen to Biddulph, 22 November 1943, 1, File 4, BFP.
65 The present author has completed a manuscript on the transfer of Movement Control technology from Britain to West Africa and South Asia during the war that will soon be submitted for publication.

Conclusion

1 Harris, *Men, Ideas, and Tanks: British Military Thought and Armoured Forces, 1903–1939*, 55.
2 Biddulph to Dorman-Smith, 28 October 1943, 1, File 4, BFP.
3 Micklem, *Transportation*, 223.
4 The British Army was successful in wedding together morale and rapid technical change because chemical warfare would clearly help beat the Germans in the last 100 days of the First World War, as Albert Palazzo has shown in his book on gas on the Western Front. Palazzo identifies a two-tiered process involving the abstract and broad commitment to change that he refers to as innovation and the detailed, material implementation of change he refers to as the process of adaptation. Palazzo. *Seeking Victory on the Western Front: The British Army and Chemical Warfare in World War I*, 4.
5 Edward M. Spiers, *Haldane: An Army Reformer* (Edinburgh: Edinburgh University Press, 1980), 12.

Bibliography

Archival sources

Anglo-Burmese Library

(www.ablmembersarea.com/)

Manuscripts

Jack Cardew, 'Extracts from the Diary of Jack Cardew, Deputy Locomotive Superintendent, BR, Insein'. http://www.ablmembersarea.com/treak-out-of-burma-military-records-html [accessed 26 April 2015].

Reginald Melville Ward Lowe, 'Some Memories of Burma and the Burma Railways'. http:// www.ablmembersarea.com/reginald-melville-ward-lowe.html [accessed 26 April 2015].

Biddulph Family Papers (BFP), London, UK

Private papers of Colonel Francis John Biddulph

[File 1] Biddulph, FJ, 'Burmah Notebook'

[File 2] 'Random Jottings'

[File 3] 'Ordnance Depot Cawnpore 1942–43'

[File 4] 'India Transport and other Burma 1942 papers'

[File 5] Records of Transportation Directorate, Burma, 1942

[File 6] 'Cuttings'

Manuscript collection, Cambridge University South Asian Archives, Cambridge, UK

Abraham Papers

Hughes Papers

Lorimer Papers

European manuscripts, British Library

RH Dorman-Smith Telegrams, Photo Eur 011

Imperial War Museum (IWM), London

Manuscript collection

R. Dorman-Smith Papers
D. Kup Papers
J U Melville Papers
E. I. Milne Papers
WH Newton Papers
H. Pearson Papers
John Perkins Papers
Stewart Papers
H. Vorley Papers
S. West Papers
W.H. Wright Papers

India Office Records (IOR), British Library

Dorman-Smith Papers, Mss Eur E215
John Rowland Papers, Mss Eur F481

Liddell Hart Centre for Military Archives (LHCMA), King's College London

G. Bomford Papers
C. M. P. Durnford Papers
C. E. M. Herbert Papers
T. J. Hutton Papers

Royal Engineers Museum Archive, Gillingham, UK

REM World War II Unit War Diaries
File No. 3, Burma Divisional Sappers and Miners (27 February 1942–31 May 1942).

File No. 13, R. Hudson. 'Diary of the Malerkotla Field Company, Indian State Forces, October 1941–May 1942'.

SOAS, University of London, Special Collections

Foucar, Draft Narrative of the First Burma Campaign December 1941–May 1942, E Coll3, C37

Official papers

India Office Records, British Library

Burma Office Annual Department Files: M/4
Burma Office Annual Files: M/3
Burma Office Collections: M/6
Burma Office Home Correspondence and Correspondence with Burma and India: M/7
Burma Office Miscellaneous Records: M/8
Burma Office 'P' (Private) and 'I' Intelligence Files: M/5
India Office Official Publications Series: V/26
India Office, Proceedings and Consultations of the GOI and its Presidencies and Provinces: P/Sec/Ind 196
India Office, War Staff Papers: LSG/11

Public Record Office (PRO), The National Archives, Kew, London

Air Ministry: TS 28
War Cabinet: CAB 106
War Office: WO 106, 172, 203, 208, 222, 373

Newspapers and periodicals

Locomotive, Railway Carriage and Wagon Review
The London Gazette
The London Times

The Mitre [Christ Church School Magazine of Perth]
Railway Gazette
Railway Gazette International
The Singapore Free Press and Mercantile Advertiser
Supplement to the London Gazette
Transactions of the Civil Engineers of Ireland

Books, articles and official publications

Alexander, Harold. *The Alexander Memoirs 1940–1945*. Edited by John North. London: Cassell, 1962.

Allen, Louis. *Burma: The Longest War 1941–1945*. London: Phoenix Press, 1984.

Anderson, E. P. 'Railway Organization in War'. *Journal of the United Services Institution* 72 (1927): 503–11.

Andrus, James Russell. *Burmese Economic Life*. Stanford: Stanford University Press, 1947.

Aung Min, Myint. *Mi-yahta Thamaing*. Yangon: Sarpay Biman, 1985.

Ausland, John E. [under pseudonym, 'The Old Gray Major']. 'Railway South of the Clouds'. *Ex-CBI Roundup* (November 1955): 12–16.

Ball, Joan. 'The Road from Mandalay'. *Blackwood's Magazine* 253, no. 1532 (June 1943): 385–93.

Ballantyne, Tony. *Orientalism and Race: Aryanism in the British Empire.* Cambridge: Cambridge University Press, 2002.

Barua, Pradeep. *The Army Officer Corps and Military Modernisation in Later Colonial India.* Hull: University of Hull Press, 1999.

Bayly, Christopher, & Harper, Tim. *Forgotten Armies: Britain's Asian Empire and the War with Japan*. London: Penguin Books, 2004.

Belden, Jack. *Retreat with Stilwell.* New York: Da Capo Press, 1975.

Bell, R. *History of the British Railways During the War, 1939–45*. London: Railway Gazette, 1946.

Bicker, Alan, Sillitoe, Paul, & Pottier, John (eds). *Essays in Development and Local Knowledge: New Approaches to Issues in Natural Resources Management, Conservation and Agriculture.* Abingdon: Psychology Press, 2004.

Biddulph, Francis John. 'Signal School at the Railway Training Centre, Royal Engineers, Longmoor'. *Railway Gazette* (2 April 1937): 656–61.

Brooke-Popham, Robert. 'Operations in the Far East, from 17th October 1940 to 27th December 1941'. *Supplement to the London Gazette* (22 January 1948): 535–76.

Brown, Ian Malcolm. 'The Evolution of the British Army's Logistical and Administrative Infrastructure and Its Influence on GHQ's Operational and Strategic

Decision-Making on the Western Front, 1914–1918'. PhD diss., King's College. London, 1996.
Bungay, Stephen. 'The Road to Mission Command: The Genesis of a Command Philosophy'. *British Army Review* 137 (2005): 22–9.
Burchett, Wilfred. *Bombs over Burma*. London: F. W. Cheshire, 1944.
'The Burma Railways During Invasion'. *Railway Magazine* 90 (July & August 1944): 215–16.
Bykofsky, Joseph, & Larson, Harold. *The Transportation Corps: Operations Overseas*. Washington, DC: US Army Center for Military History, 1957.
Campagnac, Charles Haswell. *The Autobiography of a Wanderer in England & Burma*. Raleigh, NC: Lulu Enterprises, 2011.
Campagnac-Carney, Sandra. *Burma – Memories of WWII*. Raleigh, NC: Lulu Enterprises, 2010.
Carew, Tim. *The Longest Retreat: The Burma Campaign 1942*. London: Mayflower, 1972.
Carter, Anne. *Bewitched by Burma: A Unique Insight into Burma's Complex Past*. Leicestershire: Matador, 2012.
Carter, Ernest F. *Railways in Wartime*. London: Frederick Muller, 1964.
Chan, Tak Wing, & Goldthorpe, John H. 'Class and Status: The Conceptual Distinction and its Empirical Relevance'. *American Sociological Review* 72 (2007): 512–32.
Charney, Michael W. *A History of Modern Burma*. Cambridge: Cambridge University Press, 2009.
Cheng Siok-Hwa. *The Rice Industry of Burma 1852–1940*. Kuala Lumpur: University of Malaya Press, 1968.
Christian, John LeRoy. *Burma and the Japanese Invader*. Bombay: Thacker, 1945.
Collister, Peter. *Then a Soldier*. With a foreword by General Sir Anthony Farrar Hockley. Worthing: Churchman Publishing, 1985.
Connor, W. D. *Military Railways*. Rev. edn. Washington, DC: US Government Printing Office, 1917.
Corpe, Hilda R. *Prisoner Beyond the Chindwin*. London: Arthur Barker, 1955.
Cowells, Adrian. 'Adventure on the Stilwell Road'. *Popular Mechanics* 108, no. 5 (November, 1957): 81–5; 274–6.
Croucher, Richard. *Engineers at War 1939–1945*. London: Merlin Press, 1982.
Davidson-Houston, J. V. 'Some Personal Experiences in Burma, 1942'. *Journal of the Royal United Service Institution* 88, no. 550 (1943): 121–8.
Davidson-Houston, J. V. *Armed Pilgrimage*. London: Robert Hale, 1949.
Davis, Norman. 'An Information-based Revolution in Military Affairs'. In *In Athena's Camp: Preparing for Conflict in the Information Age*, edited by John Arquilla & David Ronfeldt, 79–98. Santa Monica, CA: RAND, 1997.
Davy, G. M. O. *The Seventh and Three Enemies: The Story of World War II and the 7th Queen's Own Hussars*. Cambridge: W. Heffer & Sons, 1952.
Department of Defence. *Logistics in World War II: Final Report of the Army Service Forces*. Washington, DC: Center for Military History, 1993.

Donnisson, David. *Last of the Guardians: A Story of Burma, Britain and a Family*. Newtown: Cyhoeddwyr y Superscript, 2005.

Dorman-Smith, Reginald. 'Civil Government under Invasion Conditions'. *Journal of the United Service Institution of India* 73 (1943): 240–52.

Draper, Alfred. *Dawns Like Thunder: The Retreat from Burma 1942*. London: Leo Cooper, 1987.

Dunlop, Graham. *Military Economics, Culture and Logistics in the Burma Campaign, 1942–1945*. London: Pickering & Chatto, 2009.

Eldridge, Fred. *Wrath in Burma: The Uncensored Story of General Stilwell and International Manoeuvres in the Far East*. Garden City: Doubleday, 1946.

Enriquez, C. M. *Burma Invaded 1942*. N.p.: Margaret P. Stanford, 2012.

Fairless, Thompson. 'Standard Locomotives for Overseas Railways? Is Such an Ideal Practicable, Desirable and Economical?' *Railway Gazette* (22 June 1945): 614.

Ferguson, James. 'Anthropology and Its Evil Twin: "Development" in the Constitution of a Discipline'. In *International Development and the Social Sciences: Essays on the History and Politics of Knowledge*, edited by Frederick Cooper & Randall Packard, 150–75. Berkeley: University of California Press, 1997.

Fitzpatrick, Gerald. *No Mandalay No Maymyo (79 Survive), Unique Episodes in British History*. Lewes: The Book Guild, 2001.

Foucar, E. C. V. 'The First Burma Campaign'. *Journal of the United Service Institution of India* 72, no. 308 (1945): 59–72; no. 309 (1945): 191–202.

Foxton, P. D. *Powering War: Modern Land Force Logistics*. London: Brassey's, 1994.

Frasch, Tilman. 'Tracks in the City: Technology, Mobility and Society in Colonial Rangoon and Singapore'. *Modern Asian Studies* 46, no. 1 (2012): 97–118.

Furnivall, J. S. *Colonial Policy and Practice: A Comparative Study of Burma and Netherlands India*. New York: New York University Press, 1948.

Gardiner, R. 'The Assam Lines of Communication', *The Royal Engineers Journal* 62 (1948): 94–102.

Goodall, Felicity. *Exodus Burma: The British Escape Through the Jungles of Death 1942*. Stroad: Spellmount, 2011.

Gordon, Hampden. *The War Office*. London: Putnam, 1935.

Goss, Andrew. *The Floracrats: State-Sponsored Science and the Failure of the Enlightenment in Indonesia*. Madison: University of Wisconsin Press, 2011.

GOB. *Compendium of Governor's Acts Published in India – 1942–1945*. Rangoon: Superintendent, Government Printing, 1946.

GOI. *Report of the Indian Railway Enquiry Committee, 1937*. Delhi: Manager of Publications, 1937.

Grant, Roderick, & Cole, Christopher. *But Not in Anger: The RAF in the Transport Role*. London: Ian Allan, 1979.

Grieve, William G. *The American Military Mission to China, 1941–1942: Lend-Lease Logistics, Politics and the Tangles of Wartime Cooperation*. Jefferson, NC: McFarland, 2014.

Gwyn, P. E. C. J. 'Intelligence in the 1941–42 Burma Campaign'. *Journal of the United Service Institution of India* 77, no. 328 (1947): 524–9.

Hamilton, John Maxwell. *Journalism's Roving Eye: A History of American Foreign Reporting*. Baton Rouge: Louisiana State University Press, 2009.

Harris, J. P. *Men, Ideas and Tanks: British Military Thought and Armoured Forces, 1903–1939*. Manchester: Manchester University Press, 1995.

Harris, Murray. *Lifelines of Victory*. New York: G. P. Putnam's Sons, 1942.

Harrison, Mark. *Medicine & Victory: British Military Medicine in the Second World War*. Oxford: Oxford University Press, 2004.

Hata, Ikuhiko, Izawa, Yasuho, & Shores, Christopher (eds). *Japanese Army Fighter Aces 1931–45*. Mechanicsburg, PA: Stackpole Books, 2002.

Henderson, James H. *The Process of Military Distribution Management: A Guide to Assist Military and Civilian Logisticians in Linking Commodities and Movement*. Bloomington, IN: AuthorHouse, 2006.

Hevia, James. *The Imperial Security State: British Colonial Knowledge and Empire-Building in India*. Cambridge: Cambridge University Press, 2012.

Higham, J. B., & Knighton, E. E. (comps.). *The Second World War 1939–1945 – Army – Movements*. London: WO, 1955.

Hingston, Walter. *Never Give Up. Being Volume 5 of the History of the King's Own Yorkshire Light Infantry 1919–1942*. London: Lund Humphries, 1950.

Hnit-taya-byi Mi-yahta 1877–1977. Yangon: Hsan-thit-sa, 1977.

Hoover, Herbert. *The Memoirs of Herbert Hoover: Years of Adventure*. New York: MacMillan, 1951.

Instructions for Movement Control (Overseas in War). London: HMSO, 1938.

Jackson, Ashley. *The British Empire and the Second World War*. New York: A & C Black, 2006.

Jackson, Ashley. *Distant Drums: The Role of Colonies in British Imperial Warfare*. Brighton: Sussex Academic Press, 2010.

Johnson, N. 'A Review of the Burma Railways'. *The Engineer* (February 1949): 122–4; 150–2; 178–9; 206–8.

Jones, Paul LeRoy. 'The Withdrawal from Burma and the Stillwell Walkout'. *Ex-CBI Roundup* (May, 1992), Web Version, http://CBI-theater-3.home.comcast.net [accessed 1 August 2014].

'K'. 'Burma in My Life Time (When I was in the H.Q. Army in Burma, 1942)'. *Guardian* 9, no. 10 (October 1962): 15–16.

Keck, Stephen L. *British Burma in the New Century, 1895–1918*. Houndmills: Palgrave MacMillan, 2015.

Kerr, Ian J. *Engines of Change: the Railroads that Made India*. Westport, CT: Greenwood, 2007.

Kin Thida Oung. *A Twentieth Century Burmese Matriarch*. N.p.: Kin Thida Oung, 2007.

Kirby, S. Woodburn. *The War Against Japan*. 5 vols. London: HMSO, 1958.

Klein, Maury. *The Life & Legend of E. H. Harriman*. Chapel Hill: North Carolina University Press, 2000.Lathrop, Alan K. 'The Employment of Chinese Nationalist Troops in the First Burma Campaign'. *Journal of Southeast Asian Studies* 12, no. 2 (September 1981): 403–32.

Lathrop, Alan K. 'Dateline: Burma'. *Dartmouth Medicine* (Spring 2004): 28–35.

Law-yone, Edward. Extracts from unpublished memoir. Unpublished source gratefully provided to the author by Wendy Law-yone.

Law-yone, Wendy. *Golden Parasol: A Daughter's Memoir of Burma*. London: Vintage, 2014.

Leach, F. Burton. 'Transport Developments in Burma: Road and Air'. *Journal of the Royal Central Asian Society* 28, no. 2 (1941): 167–81.

Leigh, Michael D. *The Evacuation of Civilians from Burma: Analysing the 1942 Colonial Disaster*. London: Bloomsbury, 2014.

Lewin, Ronald. *Slim the Standardbearer: A Biography of Field-Marshal the Viscount Slim*. London: Wordsworth Editions, 1976.

Lewis, Su Lin. 'Between Orientalism and Nationalism: The Learned Society and the Making of "Southeast Asia"'. *Modern Intellectual History* 10, no. 2 (August 2013): 353–74.

Li, Laura Tyson. *Madame Chiang Kai-shek: China's Eternal First Lady*. New York: Grove Press, 2007.

Lin Yone Thit Lwin. *Ko-dwei Kwei-dab-hmat-dan*. 2 vols. Yangon: Myochit Seittak Than-ye, 1991.

Logistics in World War II: Final Report of the Army Service Forces. Washington, DC: US Army Center of Military History, 1992.

The Logistics of War: A Historical Perspective. Maxwell AFB: Air Force Logistics Management Agency, 2000.

Lourie, George E. 'Development of a Military Railway Service'. *Military Review* 26, no. 6 (September 1946): 26–33.

Ludu Ù Hlá, *Thadìn-sa-myà Pyaw-pyá-deì Sit-twìn Bama-pyi*. 4 vols. Mandalay: Ludu Press, 1968–9.

Lunt, James. 'A Hell of a Licking: Some Reflections on the Retreat from Burma, December 1941–May 1942'. *Journal of the Royal United Services Institution* 130, no. 3 (1985): 55–8.

Lunt, James. *The Retreat in Burma 1941–1942*. Trowbridge: David & Charles, 1989.

Lyall-Grant, Ian. 'The British View: British Railway Demolitions in Burma'. In *Railwaymen in the War: Tales by Japanese Railway Soldiers in Burma and Thailand 1941–47*, edited by Kazuo Tamayama, 74–6. London: Palgrave Macmillan, 2005.

Lyall-Grant, Ian, & Tamayama, Kazuo. *Burma 1942: The Japanese Invasion, Both Sides Tell the Story of a Savage Jungle War*. Chichester: Zampi Press, 1999.

MacKay, James Noble. *A History of the 4th Prince of Wales's Own Gurkha Rifles*. Edinburgh: William Blackwood & Sons, 1952.

MacKay, James Noble. *History of the 7th Duke of Edinburgh's Own Gurkha Rifles*. Edinburgh: William Blackwood & Sons, 1962.

McKenna, Frank. *The Railway Workers 1840–1970*. London: Faber and Faber, 1980.

MacLachlan, W. P. G. 'Burma Journey'. *The National Review* 120, no. 724 (June 1943): 475–82; 121, no. 725 (July 1943): 61–7, 140–5.

McCrum, Ronald. *The Men Who Lost Singapore*. Singapore: Singapore University Press, 2017.

McGlade, Fred. *The History of the British Army Film & Photographic Unit in the Second World War*. Solihull: Helion, 2010.

McLynn, Frank. *The Burma Campaign: Disaster into Triumph 1942–45*. New Haven, CT: Yale University Press, 2011.

Magruder, Carter B. *Recurring Logistic Problems as I Have Observed Them*. Washington, DC: Center for Military History, 1991.

Mains, A. A. *The Retreat from Burma: An Intelligence Officer's Personal Story*. New Delhi: Palit & Palit, 1973.

Mains, A. A. *A Soldier with Railways*. Chippenham: Picton, 1994.

Majumdar, B. N. *Administration in the Burma Campaign (1941–1945)*. Delhi: Clifton, 1952.

Manton, Lionel. 'Communication'. *Journal of the Institute of Transport* 19 (1938): 285.

Marston, Daniel P. *Phoenix from the Ashes: The Indian Army in the Burma Campaign*. Westport, CT: Praeger, 2003.

McKenna, Frank. *The Railway Workers 1840–1970*. London: Faber and Faber, 1980.

Maybury, Maurice. *Heaven-Born in Burma*. Vol. 2. *Flight of the Heaven-Born*. Somerset: Folio Hadspen, 1985.

Metcalf, Thomas R. *Imperial Connections: India in the Indian Ocean Arena, 1860–1920*. Berkeley: University of California Press, 2007.

Micklem, R. (comp.). *The Second World War 1939–1945 Army: Transportation*. London: WO, 1950.

Military Railways: Rule Book for Use by Indian Troops in India and Other Eastern Countries. Moghalpura: North Western Railway Press, 1944.

Millem, Fred. 'Rangoon Battalion', introduced by Patricia Pringle. http://www.bbc.co.uk/history/ww2peopleswar/categories/c55521/ [accessed 9 September 2017].

Mitchell, Vic, & Smith, Keith. *Branch Lines to Longmoor*. Midhurst: Middleton Press, 1987.

Mole, Robert. *The Temple Bells Are Calling: A Personal Record of the Last Years of British Rule in Burma*. London: Pentland Books, 2001.

Moore, Aaron. *Constructing East Asia: Technology, Ideology, and Empire in Japan's Wartime Era, 1931–1945*. Stanford: Stanford University Press, 2013.

Moore, David M., Bradford, Jeffrey P., & Antill, Peter D. *Learning from Past Defence Logistics Experience: Is What Is Past Prologue?* London: The Royal United Services Institute for Defence Studies, 2000.

Moreman, Tim. *The Jungle, the Japanese and the British Commonwealth Armies at War, 1941–1945*. London: Frank Cass, 2005.

Munro, H. P. 'Retreat December 1941 to May 1942: 24 (RB) Field Company and Malerkotla Field Company'. In *The Royal Bombay Sappers & Miners, 1939–1947*, edited by Laurie & Cecilia Norman, 271–82. Bombay: Royal Bombay Sappers & Miners Officers' Association, 1999.

Myint Swei. 'Myanma-pyi Mi-yahta Hnit-taya Thamaing'. *Thwei-thauk Magazine* 318 (June, 1972): 148–9.

Naono, Atsuko. *State of Vaccination: The Fight Against Smallpox in Colonial Burma.* Himayatnagar: Orient BlackSwan, 2009.

Nemoto, Kei (ed.). *Reconsidering the Japanese Military Occupation in Burma (1942-45).* Tokyo: Research Institute for Languages and Cultures of Asia and Africa (ILCAA), Tokyo University of Foreign Studies, 2007.

Okell, John. *A Guide to the Romanization of Burmese.* London: Royal Asiatic Society of Great Britain and Ireland, 1971.

Olmstead, Joseph A. *Leading Groups in Stressful Times: Teams, Work Units, and Task Forces.* Westport: Quorum Books, 2002.

Pakenham-Walsh, R. P. *History of the Corps of Royal Engineers.* Vol. IX. Chatham: Institution of Royal Engineers, 1958.

Pal, Dharm. *Campaign in Western Asia.* Bombay: Combined Inter-services Historical Section, India & Pakistan, 1957.

Palazzo, Albert. *Seeking Victory on the Western Front: The British Army and Chemical Warfare in World War I.* Lincoln: University of Nebraska Press, 2000.

Perry, F. W. *The Commonwealth Armies: Manpower and Organisation in Two World Wars.* Manchester: Manchester University Press, 1988.

Porter, G. T. 'The Lines Behind the Lines in Burma'. *Railway Magazine* 89, no. 548 (November & December 1943): 325–7.

Prasad, Bisheshwar (ed.). *The Retreat from Burma 1941–42.* Official History of the Indian Armed Forces in the Second World War 1939–45. Delhi: Combined Inter-Services Historical Section (India & Pakistan), 1952.

Prendergast, W. H. 'A Galway Engineer in Assam'. http://places.galwaylibrary.ie/history/chapter70.html [accessed 28 November 2014].

Pugsley, Christopher. 'We Have Been Here Before: The Evolution of the Doctrine of Decentralised Command in the British Army, 1905–1989'. *Sandhurst Occasional Papers* no. 9 (2011): 1–39.

The Railway Handbook 1945–1946. London: The Railway Publishing Company, 1945.

Raina, B. L. (ed.). *Campaigns in the Eastern Theatre.* Official History of the Indian Armed Forces in the Second World War 1939–45. Medical Services. New Delhi: Combined Inter-Services Historical Section, India & Pakistan, 1964.

A Record of Large-Scale Organisation and Management, 1923–1946. London: London Midland and Scottish Railway Company, 1946.

Robertson, Joan. *Maymyo More Far: A Walk Out of Burma 1942.* Banbury: Norman Hudson, 1999.

Rogers, Everett M. *Diffusion of Innovations.* 4th edn. New York: The Free Pess, 1995.

Romanus, Charles F. *Stilwell's Mission to China.* Washington, DC: US Army Center for Military History, 1987.

Ronald, David W., & Christensen, Mike. *The Longmoor Military Railway: A New History*. Vol. 1. *1903–1909*. Lydney: Lightmoor Press, 2012.
Rose, Patrick. 'Allies at War: British and US Army Command Culture in the Italian Campaign, 1943–1944', *Journal of Strategic Studies* 36, no. 1 (2013): 42–75.
The Royal Bombay Sappers & Miners 1939 to 1947. Bombay: Royal Bombay Sappers & Miners Officers' Association, 1999.
Rupenthal, Roland G. *Logistical Support of the Armies*. 2 vols. Washington, DC: Center for Military History, 1995.
Russell, S. Farrant. *Muddy Exodus: A Story of the Evacuation of Burma, May, 1942*. London: Epworth Press, 1944.
Saha, Jonathan. *Law, Disorder and the Colonial State: Corruption in Burma c. 1900*. Houndmills: Palgrave Macmillan, 2013.
Saito, Zyoichi. 'Diary of a Company Commander'. In *Railwaymen in the War: Tales by Japanese Railway Soldiers in Burma and Thailand 1941–47*, edited and translated by Kazuo Tamayama, 69–73. London: Palgrave Macmillan, 2005.
Sandes, Edward Warren Caulfield. *The Indian Sappers & Miners*. Kirkee: Institution of Royal Engineers, 1948.
Sandes, Edward Warren Caulfield. *From Pyramid to Pagoda: The Story of the West Yorkshire Regiment (The Prince of Wales's Own) in the War, 1939–45 and Afterwards*. With a foreword by Field-Marshal Sir William Slim. London: F. J. Parons, 1952.
Sandes, Edward Warren Caulfield. *Indian Engineers, 1939–1947*. Kirkee: Institution of Military Engineers, 1956.
Scott, D. G. Lacy. 'Burma: The Great Withdrawal'. *The Burma Campaign Society Newsletter* 15 (March, 2010): 13–15.
Scott, James C. *Seeing Like a State: How Certain Schemes to Improve the Human Condition Have Failed*. New Haven: Yale University Press, 1998.
Seagrave, Gordon S. *Burma Surgeon*. New York: W. W. Norton, 1943.
Seekins, Donald M. *Burma and Japan since 1940: From 'Co-Prosperity' to 'Quiet Dialogue'*. Copenhagen: NIAS, 2007.
Sein Tin, Ù [Theippan Maung Wa]. *Sit-atwìn Neí-sin Hmat-tàn*. Rangoon: Pugan Sa-oub, 1966.
Sheffield, G. D. 'The Australians at Pozieres: Command and Control on the Somme, 1916'. In *British General Staff: Reform and Innovation*, edited by David French & Brian Holden Reid, 94–105. London: Routledge, 2004.
Shores, Christopher, Cull, Brian, & Izawa, Yasuho. *Bloody Shambles. Vol. 1. The Drift to War to the Fall of Singapore*. London: Grub Street, 1992.
Shores, Christopher, Cull, Brian, & Izawa, Yasuho. *Bloody Shambles. Vol. 2. The Defence of Sumatra to the Fall of Burma*. London: Grub Street, 1993.
Short, Stanley W. *On Burma's Eastern Frontier*. With a foreword by S. H. Middleton West. London: Marshall, Morgan & Scott, 1945.
Slater, Robert. *Guns through the Arcady: Burma and the Burma Road*. Madras: Diocesan Press, 1943.

Slim, William. *Defeat into Victory*. London: Cassell, 1956.

Smith, Benjamin R. '"Indigenous" and "Scientific" Knowledge in the Central Cape York Peninsula'. In *Local Science vs. Global Science: Approaches to Indigenous Knowledge in International Development*, edited by Paul Sillitoe, 75–90. New York: Berghahn Books, 2013.

Spiers, Edward M. *Haldane: An Army Reformer*. Edinburgh: Edinburgh University Press, 1980.

Stevenson, D. F. 'Despatch on Air Operations in Burma and the Bay of Bengal Covering the Period January 1st to May 22nd, 1942'. *Second Supplement to the London Gazette* (5 March 1948): 1713–33.

Stewart, Adrian. *The Underrated Enemy: Britain's War with Japan, December 1941–May 1942*. London: William Kimber, 1987.

Stewart, Keith G. 'The Evolution of Command Approach'. Toronto: Defence Research and Development Canada, n.d., 1–12.

Stone, H. David, Jr., *Vital Rails: The Charleston & Savannah Railroad and the Civil War in Coastal South Carolina*. Columbia: University of South Carolina Press, 2008.

Stowe, Leland. *They Shall Not Sleep*. New York: A.A. Knopf, 1944.

Strangleman, Tim. 'Railway and Grade: the Historical Construction of Contemporary Identities'. PhD diss., University of Durham, Durham, 1998.

Tamayama, Kazuo (comp.). *Railwaymen in the War: Tales by Japanese Railway Soldiers in Burma and Thailand 1941–47*. London: Palgrave MacMillan, 2005.

Theìn Hpe Myínt. *Sit-atwìn Hkayì-thi*. Rangoon: Pugan Sa-oub, 1966.

Thomas, John. *Line of Communication: Railway to Victory in the East*. London: The Locomotive Publishing Company, 1947.

Thompson, H. L. *The Official History of New Zealand in World War II. New Zealanders with the Royal Air Force Vol. III*. Wellington: Historical Publications Branch, 1959.

Thompson, Julian. *Lifeblood of War: Logistics in Armed Conflict*. London: Brassey's, 1991.

Thompson, T. O. 'Burma Retreat 1942'. *Journal of the Royal Army Medical Corps* 90, no. 1 (January 1948): 23–9; 90, no. 2 (February 1948): 46–60.

Tin Maung Aye. *It Started in Phyu: A Childhood Memoir from World War II in Burma and the Journey Forward*. Yangon: Myanmar Book Centre, 2013.

Universal Directory of Railway Officials and Railway Year Book 1941–1942. London: The Directory Publishing Company, 1941.

Van Creveld, Martin. *Supplying War: Logistics from Wallenstein to Patton*. 2nd edn. Cambridge: Cambridge University Press, 2004.

Verma, S., & Anand, Vijay Kumar. *The Corps of Indian Engineers, 1939–1947*. Delhi: Historical Section, Ministry of Defence, GOI, 1974.

Vorley, J. S., & Vorley, H. M. *The Road from Mandalay*. Hernes Keep: Wilton 65, 2002.

Wagg, Alfred. *A Million Died! A Story of War in the Far East*. London: Nicholson & Watson, 1943.

War Office. *Military Engineering (vol. III) Railways*. His Majesty's Stationery Office, 1929.

War Office. *Military Railways Rule Book*. London: HMSO, 1938.

War Office. *Notes on Military Railway Engineering Part I (Survey)*. London: The QMG, 1940.

War Office. *Notes on Military Railway Engineering Part II Engineer*ing. London: The QMG, 1940.

War Office. *Notes on Military Railway Engineering Part IV: Operating*. London: The QMG, 1942.

Warren, Alan. *Burma 1942: The Road from Rangoon to Mandalay*. London: Continuum, 2011.

Weber, Max. *Economy and Society: An Outline of Interpretive Sociology*. Berkeley: University of California Press, 1978.

Wedgwood, Ralph L. 'Paper by Ralph L. Wedgwood'. *Railway News* 84 (1905): 789.

Wetzel, Nevin F. *Engineer in the Far East: Memoirs of World War II in China, Burma, and India*. Denver, Colorado: Nevin F. Wetzel, 1984.

Wilkie, H. G. *Report on the Effect of Denial and Demolition Schemes in Burma*. Alipore: Bengal Government Press, 1942.

Williams, G. *Citizen Soldiers of the Royal Engineers Transportation and Movements and the Royal Army Service Corps, 1859–1965*. London: Institute of the Royal Corps of Transport, 1969.

Willmott, H. P. *When Men Lost Faith in Reason: Reflections on War and Society in the Twentieth Century*. London: Greenwood, 2002.

Wolmar, Christian. *Engines of War: How Wars Were Won & Lost on the Railways*. London: Atlantic Books, 2010.

Woods, Philip Graham. *Reporting Retreat: War Correspondents in Burma*. London: Hurst. 2017.

Index

Abraham, W. E. V. (Maj. Gen.) 159–60
Air, William J. 38, 39, 84
air raids 4–5, 23, 53–4, 56, 58, 65, 69, 75–6, 90, 92, 101–2, 105, 107, 113, 114, 117–18, 119, 120, 123, 124, 126, 128–9, 135, 137–9, 151, 166, 173
Akyab (Sittwe) 89
Alexander, Harold R. L. G. (Gen.) 87, 89, 94, 129, 131, 138, 144, 148, 155, 158
ambulance trains 100, 119, 152, 207 n.13
American Army Forces 130, 150
American-British-Dutch-Australia Command (ABDACOM) 69
American Chinese Liaison Mission (CLM) 108, 111
American Volunteer Group 75
Amingaon 175
Anderson, E. P. 37
anti-aircraft cover and units 68, 167
Appadurai, Arjun 2
Arakan 89, 164
armoured trains 68
Army Technical Centre (Jullundur) 170
Arsenal Transport Company 39
Assam 2, 10, 139, 155–9, 172–4
Assam Bengal Railway 173
Assam Lines of Communication 10, 159, 171–8, 186
Australia 26, 35, 47, 69, 169
Ava Bridge 84, 131, 135, 142–5, 150–1, 154
Aye Maung, U. 121

Ba Gale, U 121
Baker, H. C. 128
Ballantyne, Tony 187 n.4 (*see also* knowledge and learning)
Bangkok 2
Basra 27, 29
Bassein 44, 122
  as an alternative to Rangoon 89
Bassein Branch 44
Bay of Martaban 62, 68
Beijing 24
Belgium 23
Bengal 22
Bengal and Assam Railway 29–30, 172–3
Beyer-Garratt locomotives 113, 142
Biddulph, Francis John (Col./Brig.) x, 7–13, 15–19, 22–30, 33, 49, 51–2, 57–62, 67–91, 94, 97, 99, 101, 103, 107, 108, 110, 112, 131, 134, 137–9, 144, 147, 149, 151, 153–72, 178, 181–5
Bilin River 63–4
Binnie, G. H. 92, 171
Blakeney, C. N. 82, 171
Bombay 28–9
The Boxer Rebellion 24
Brahmaputra River 174–5
Brewitt, Charles Patrick 38–9, 47, 60, 81–2, 86, 88, 91–2, 99–101, 103, 106, 108, 110, 112, 114, 117–18, 151, 155, 169, 171–2
Britain 1–3, 7, 10–11, 14, 16–30, 34, 35, 38–42, 45, 48–9, 51–2, 54, 58–9, 64–5, 69, 105, 119, 158, 179, 181, 184, 186
British Library 161
Brown, N. W. G. (Capt.) 58, 108
Bryant, S. C. 82, 169
Burcorps (I Burma Corps) 16, 89, 155
Burma Army 6, 48, 53, 55–6, 60, 62–5, 70, 73, 75, 77, 80–1, 85–7, 90, 91, 103, 162, 170
Burma Auxiliary Force 55–7, 61, 81
Burma, Government of (GOB) 51, 56, 90, 111, 138, 140, 161, 163, 169–71
Burma Military Police 137
Burma Mines Railway 36, 44, 134
Burma Mining Corporation Railway (*see* Burma Mines Railway)
Burma Railways Battalion, Burma Auxiliary Force 55–7, 61
Burma Rifles 126, 137

Burma Road 2, 108, 163
Burmah Oil Company 66, 84
Burmascape 2–3
Busher, R. A. (Capt.) 30, 92

Calcutta 25, 27, 97, 165, 173–4
Cant, R. D. 171
Cardew, Jack 85, 92, 137, 169
Carr, Raymond Taylor (Lt.) 98, 106, 108, 110, 120, 129
Carson, (Sir) F. (Brig.) 28–9
Chabua 175
Chance, William Henry 13, 47, 56, 58–62, 73, 75–81, 84, 91, 99, 100, 110, 122
Chandranathpur 175
Chiang Kai-shek 93, 108, 119
Chiang Mei-ling 93
Chiengmai 85
China 2, 24, 52, 54, 69, 83, 88, 94, 107–8, 111, 118, 130, 135, 146, 150, 163
Chindwin River 149, 155
Chinese Expeditionary Force 89, 98, 104, 107–31, 137, 140, 141–6, 148–52, 161, 168 (*see also* 22nd Chinese Division; 96th Chinese Division)
  conflict with British and imperial forces 207 n.13
Chittagong 175
Christensen, Mike 22
Chungking 111, 130
coal and coaling 54, 69, 75, 112–13, 118, 126–7, 136, 174, 176
Cochrane, (Sir) Archibald (Gov.) 57
Cologne 22
command philosophy 98–9
communications 4, 5, 13, 41, 66, 67, 69, 87, 100, 106, 107, 114, 118, 120, 128, 131, 135–6, 147, 154, 158, 172
  telegraphic 84, 97, 151–2, 154
  telephonic 67, 74, 79, 84, 97, 100, 102, 114, 173
Constantinople 39
Copeman, A. N. 171
Coward, Duncan (Maj.) 98–104, 108, 114–29, 145
Crewes Mechanics Institution 39

Daiku 121
Davidson-Houston, James Vivian (Col.) 119
denial and demolition schemes 63, 87–8, 109, 111, 144–5
Dennys, L. E. (Maj. Gen.) 111
Deolali 13, 29, 148
Department of War Transport 165, 169
Derby 22, 23
Digboi 174
Dimapur 175
Directorate of Movements and Quartering 16
Donnisson, Frank 149, 150, 162
Dorman-Smith, Reginald (Governor) 51, 69, 79, 82, 137–9, 141, 143–5, 154, 155, 182
  complaints about behaviour of Chinese troops 142
  criticisms of 54, 162–3
  and militarization of civilian staff 39, 76, 80–1, 82, 90–1, 161–2, 168
  preparations for the war 55, 184
Dulwich 22
Dunkirk 23, 158
Dunlop, Graham 9, 159
Durrell, W. H. 48
Dutch East Indies 7

East Bengal Railway 173
East Indian Railway Engineers 106
Edwards, C. C. R. 82, 169
Egypt 18, 22
engineers 36, 37, 39, 146, 171, 178, 185, 207 n.13
  American 37
  Chinese 109
  see also Indian Engineers; Royal Engineers
The Enlightenment 7
Enriquez, Colin Metcalfe Dallas (Maj.) 93
Ethiopia 18
evacuation (civil) 1, 4, 30, 47, 60, 74–5, 79, 85, 88, 91, 100, 102, 112, 117, 119, 121, 129, 130, 138–42, 145, 148, 155, 157, 160, 162–3, 166, 170–1, 181, 185 (*see also* refugees)
  of Maymyo 140–1
  of Rangoon 69, 74–5, 76, 79, 84, 85–7, 139, 140

Federated Malay States Railways 2
Ferguson, James Alexander 81, 170

5th Army (Chinese) 99
50th Field Park Company, Sappers & Miners 104
1st Burma Division 88–9, 98–109, 143
First World War 2, 3, 17–24, 29, 34–5, 39, 61, 64, 98, 181, 211 n.4
Fitzherbert, A. F. 81, 171
Foucar, E. C. V. 54, 76, 160–2, 168
France 22, 203 n.77
Franco-Prussian War (1870) 203 n.77
Furnivall, J. S. 3

Gardiner, R. 56, 59, 159, 162
  replacement of Lionel Manton in India 208 n.12
Gauhati 173, 176
Geddes, Eric 181
Germany 17, 18, 24, 48 (*see* Prussia)
  German Army 98
Ghana (*see* Gold Coast)
Goddard, Eric Norman (Maj. Gen.) 57, 60, 61, 76–7, 80, 89, 133, 138, 175
  on militarization of civil transport in Burma 61, 77
  on situation of Rangoon as a base installation 133
  on subject of Transportation Technique 60
Godwin-Austen, A. R. (Lt.-Gen.) 177
Golakganj 175
Gold Coast (Ghana) 33, 158
Goss, Andrew 7
Government of Burma Act 1935 90–1
The Great Western Railway 18
The Grouping Act 18

Haig, (Sir) Douglas (Field Marshal) 181
Haldane, Viscount Richard Burdon 183
Hancock, J. 171
Harris, J. P. 11
Harrison, Mark 11
Henzada 43, 44
Henzada-Myanaung Branch Line (*see* Myanaung Branch)
Hevia, James 15
Hitler, Adolph 18
Hodkinsons 22
Hoover, Herbert 37
Hosie (Lt.) 117, 127
hospitals 38, 148, 152, 157, 169
Howes, E. D. G. 171
Hudson, E. R. B. 'Roy' (Maj.) 105
Hutton, Thomas J. (Gen.) 69, 87, 89, 90, 137

Imperial Japanese Army 33, 126, 173
Imperial Japanese Army Air Force 4, 53, 58, 65, 80, 92, 102, 104, 113–17, 128, 135, 139, 146, 152, 165, 166
Imperial War Museum 9, 161
India 1–11, 13, 14, 15, 16, 20, 22, 24–9, 34, 35, 40, 41, 45, 48, 51, 53, 55–60, 73, 74–7, 80, 86, 91, 111, 112, 113, 124, 130, 138, 146, 147–8, 152, 155, 156, 157–79, 181–6
Indian Army 10–11, 14, 15, 16, 24, 26, 55, 57, 119, 162, 163, 165, 169, 170, 174–5, 178
Indian Engineers (IE) 10, 26, 27, 29, 30, 58, 74–7, 86, 91, 106, 108, 117, 120, 170, 171
Indian Railways 45, 171
Indian Railways Enquiry Committee (*see* Wedgwood Committee)
Inland Water Transport (IWT) 25, 27, 29, 37, 60, 61, 77, 80, 137, 167
Insein Workshops 43, 82, 84–6, 113, 148
Inter-Allied Railway Sub-Commission 22
Iraq 7, 15, 16, 20–2, 25–30, 51, 57, 74, 89, 157, 158, 167, 174
Irrawaddy Delta 73
Irrawaddy Division 97
Irrawaddy Flotilla Company 53, 78, 140, 156
Irrawaddy River 66, 70, 84, 89, 90, 101, 129, 135, 139, 142–6, 148, 152
Irrawaddy Valley 12, 71, 87, 89, 108
Italy 18, 23, 98, 177

Japan 1, 3, 4, 18, 20, 48, 52–6, 73, 99, 124
Japanese capture of Singapore 70
Japanese invasion of Burma 4, 8, 11–12, 30, 34, 46, 51–7, 60–6, 68, 69, 70, 71, 73–7, 85, 87–90, 94, 97, 99–105, 111, 112, 120, 123, 125, 128–30, 142, 144, 145, 148, 149, 150, 152, 155, 160, 161, 164, 166, 172, 173, 176, 181
Japanese invasion of India 10–11
Japanese invasion of Southeast Asia 3
Johnson, Arthur 81, 145

Johnson, Norman 35, 81, 84, 91
Jones, Paul Leroy 'Casey Jones' (Capt.) 130, 151
Jorhat 175
Jullundur 26–30, 77, 170–2

Kalewa 148, 151, 155
Kambe 60
Kanbya Station 149
Kapur, D. N. 171
Katha Branch 44
Katha-Naba Branch Line (*see* Katha Branch)
KawKareik Pass 62
Keck, Stephen L. 2
Kirby, S. Woodburn 60, 62, 182
Kirkee 25
knowledge and learning 35–7
  imperial circulation of 6, 14, 21–30, 94–5, 184, 187 n.4
  knowledge communities and 184
  local knowledge 35–9
Kyaukpadaung Branch 44, 112–13, 123, 165
Kyaukse 100, 108–9, 115, 129, 148
Kyaukto 64
Kyaw Zan, U 121
Kyidaunggan 120

Lang, E. C. (Brig.) 16
Lashio 59, 99, 100, 145
Lashio Branch 37, 44, 108, 110–12, 133–5, 139, 141–2, 144
Law-yone, Edward 91–3, 143
Law-yone, Wendy 220
League of Nations 18
Ledo Road 175
Leigh, Michael 187 n.1
Lekhhapani 175
Letpadan 88
Letpadan-Bassein Branch Line (*see* Bassein Branch)
Lewe 128
Lewis, John Douglas 81, 98–107, 110–11, 169–70
Liddell-Hart Collection (Kings) 9
Locomotive Magazine 20
locomotives 4, 27, 28, 34, 62–7, 85, 112, 124, 126, 134, 142, 145, 148, 150–1, 154, 164–5, 174–5
logistics (*see* military logistics)
The London and North Western Railway 18, 39
The London, Midland and Scottish Railway (LMSR) 18
Longmoor Railway Training Institute 7, 8, 10, 12, 17–24
  Longmoor training as acquired at 73, 77, 157–8, 172, 183, 186
Longmoor Railway Training Institute 7–8, 10, 12, 16–24, 73, 77, 146, 172, 183, 186, 187 n.9
Lumding 166, 174–6
Lunt, James 86, 103
Luo, Zhuoying (Gen.) 150–1
Lusk, James Wallace (Dr.) 152

McAllister, N. 171
McCrum, Ronald 187 n.1
MacDougall, Raibert M. 53
McKenna, Frank 38–9
McMullen, D. J. (Maj. Gen.) 165
Madaya Railway 44
Madras 74
Magwe Aerodrome 75
Mahlwagon 88
Mains, Anthony 'Tony' 176
Malaria 155, 157, 169, 175
Malaya 52, 170, 177
Malayan Command 52
Malerkotla Field Company, Sappers & Miners 104–5, 109
Mallet locomotives 113, 142
Manchukuo 36
Mandalay 43, 70, 71, 75, 82–93, 97–102, 112, 114, 118, 121, 125, 129, 133, 135–44, 146, 166
Mandalay-Madaya Light Railway (*see* Madaya Railway)
Mandalay-Maymyo-Lashio Branch Line (*see* Lashio Branch)
Mandalay-Myitkyina Branch Line (*see* Myitkyina Branch)
Mandalay railway yard 166
Mandalay Station 93
Mandalay-Ye-U Section (*see* Ye-U Branch)
Manifold, P. 38
Manipur 150
Manipur-Lumding Section 166
Manipur Road 166, 174–5

Manipur Road Station 174
Manoeuvre Warfare 98
Manton, Chris 189 n.21, 208 n.12
Manton, Lionel (Col.) 7, 9, 13, 22–4, 29, 56, 74, 81, 94, 147–8, 157, 159, 162, 184, 189 n.21
  resignation and departure from India 208 n.12
Mariani 175
Markwick, E. J. R. 171
Marston, Daniel P. 9
Martaban 62–5, 68, 73
Martaban Branch 44, 67–9
  loss of 65
Martin, J. C. (Brig.) 111, 114–15, 149–50
Maung Ba Win 127
Maung Than Pe 127
Maymyo 84, 87, 90, 116, 135, 139–46
Maymyo Evacuation Special 140–1
Meiktila 98, 100, 104, 127–8
Meiktila Station 127
Metcalf, Thomas R. 26
Meza 144
Micklem, R. (Brig.) 182
Middle East 4, 15, 24–7, 51, 157, 159
military-civil relations 27, 29, 51–65, 73, 75–81, 88–91, 100, 102–6, 109–11, 119, 120, 122, 163, 164–7, 187 n.1
  and militarization 2, 5–6, 12, 13, 39, 51, 55–6, 61–2, 73–95, 147, 157–79, 185
military adaptation and innovation 3, 6–8, 11, 12, 14, 15, 21–2, 30–1, 45, 51–3, 73–4, 157, 164–5, 168–79, 181–5, 211 n.4
  the "defeat paradigm" as catalyst for 8–9, 164–5, 181
  in the First World War 211 n.4
  role of reformers in implementing technological change 3–5, 7–11, 14, 15, 30–1, 41–2, 49, 137, 157–9, 165, 178, 181–3
military logistics 15, 16, 18–21, 58, 181
  and First Burma Campaign 9, 119, 159
  Forward Area 103–4, 115, 117
  Forward Support 12, 52, 62–3, 65, 67, 71, 130
  and India 10, 16, 26, 158–9
  Rear Area (*see also* Movement Control; Transportation Technique)
Milne, E. I. 47, 56, 60, 82, 84, 91–2, 110, 118, 142, 144
Mingaladon 69, 88
Mingaladon Branch 88
Mingaladon Cantonment Branch Line (*see* Mingaladon Banch)
Mogaung 144, 146, 159
Monywa 139–40, 149
Monywa Station 140
Moore, Aaron 36
Moreman, Tim 9
Morgan, C. H. 171
Morton, John 58
Moulmein 62–3
Moulmein-Ye Section (*see* Ye Branch)
Movement Control 14, 18, 20, 23, 31, 142, 158, 174, 177–9, 186
Mu Valley State Railway (*see* Myitkyina Branch)
Myanaung Branch 44
Myingyan 114
Myingyan Branch 44
Myingyan Station 127
Myitnge Carriage and Wagon Workshops 82, 85–6, 89, 82, 102, 113, 129, 135–8, 143–8
Myitkyina 75, 139, 140, 145, 148, 150, 154, 172
Myitkyina Aerodrome 141
Myitkyina Branch 13, 43, 44, 110, 122, 131–5, 144, 147–56
Myohaung Junction 92–3, 118, 123, 129, 135–6, 139–45, 151

Naono, Atsuko 97
New Delhi 13, 59, 74, 147
New York Stock Exchange 18
Newton, W. H. 107
Nigeria 182
96th Chinese Division 120
No. 1 Transportation Training Centre 27, 29
No. 2 Engineer Depot (*see* No. 2 Transportation Training Centre)
No. 2 Transportation Training Centre 29, 170–2
No. 3 Railway Construction & Maintenance Group, (Royal) Indian Engineers 74–5, 79, 86, 90, 92, 172

No. 5 Engineer Depot, Indian Engineers (*see* No. 1 Transportation Training Centre)
No. 8 Railway Operating Group 17
Normandy 23
North Africa 22
North Western Railway 22, 35

Olmstead, Joseph A. 38
139 Railway Operating Company, (Royal) Indian Engineers 106, 117, 120
165 Indian Railway Operating Company (Burma) 170
168 Indian Railway Construction Company (Burma) 170
170 Indian Railway Operating Company (Burma) 170–1
172 Indian Railway Construction Company (Burma) 170
172 Indian Railway Workshop Company (Burma) 170
174 Indian Transportation Stores Company (Burma) 170

PaiForce 10
Palazzo, Albert 11
Palmer 102–3
Pandu 175–6
Parbatipur 174–5
Parker, Thomas C. 35
Pegu 43, 60, 63, 65–71, 79, 83, 84, 87–9
Pegu Force 67
Pegu-Martaban Branch Line (*see* Martaban Branch)
Pegu River 67–70
Pegu-Thongwa Branch Line (*see* Thongwa Branch)
Peking Relief Force 24
Penwegon 121
Perth 35
Peters, George 114, 117
Powell, M. 39
Prendergast, William Henry 29–30, 74–5, 81, 86, 88, 90–3, 106–7, 134, 153, 155, 157, 172–3
Proctor, Ernest 39, 81, 84, 91, 169
Prome 66, 88, 106
Prome Branch 44, 73–95
Prussia 203 n.77
Pugsley, Christopher 98
Pyawbwe 100, 104, 108, 114, 117, 119, 127–8, 194
Pyinmana Junction 102, 104, 107, 108, 111–23, 128, 169
Pyinmana-Kyaukpadaung Branch Line (*see* Kyaukpadaung Branch)
Pyu 99, 109, 110
Pyuntaza 103, 121

Quebec Conference 173, 175

race and racial divisions 2, 3, 5, 8, 11–12, 33–9, 42–9, 122–3, 161, 185
Railway Administration, departmental vs divisional system 41–2
Railway Clearing House 18
railway culture 59
railwaymen, colonial 4–5, 33–49, 57, 63–8, 185
The Railways Act of 1921 (*see* The Grouping Act)
Rangoon 2, 5, 8, 13, 35, 42, 43, 48, 52–118, 121–4, 133–5, 139–46, 151, 159, 165–6, 171–2, 185
Rangoon and Irrawaddy State Railway (*see* Prome Branch)
Rangoon-Mandalay Trunk Line 63, 67–70, 87–8, 99, 103, 106, 109, 121, 135, 142
Rangoon-Prome Branch Line 44
refugees 53, 75, 79, 86, 102, 105, 117–19, 126, 140, 141, 143, 149–53
retrenchment 12, 22, 34, 45–8 (*see also* Wedgwood Committee)
Rhine Army 22
Rogers, A. B. (Maj.) 58
rolling stock 2, 23, 27, 41, 63, 57, 109, 117, 124, 140, 150, 174, 175
Ronald, David 22
Rose, Patrick 98
Rowland, (Sir) John 46–7, 55–6, 59, 76, 137–40, 142, 144, 155, 160, 162
Royal Air Force 75, 88, 166
Royal Engineers 7, 11, 48, 183
  in the FBC 65, 73 (*see also* Lionel Manton; Francis John Biddulph)
  training as railway engineers 16–31

sabotage 5, 120–5, 126, 176, 203 n.77
Sagaing 140, 144
Saha, Jonathan 97

Saigon 2
St. Luke's Church 110
Salonika 39, 40
Salween Estuary 62–3, 66
Samon 129
San Paw, U 65
Sandes, E. W. C. 24
Santahar 175
Saw, U (Prime Minister of Burma) 121
School of Military Engineering 22
School of Oriental and African Studies 161
Scott, D. G. Lacey (Capt.) 64
Scott, James Bruce (Maj. Gen.) 99, 101, 102–5, 107
Scott, James C. 36
2nd Burma Brigade 86, 105
2nd Kings Own Yorkshire Light Infantry Regiment 64, 150
2nd Tank Regiment 70
Second Sino-Japanese War 108, 111
Serbia 40
Serbian Army 39
7th Armoured Brigade 89
7th Burma Rifles 100
7th Hussars 70
7 Rajputana Rifles 5
17th Indian Division 64–9, 79, 89, 130–1
Shah, Nk Sardar Ali 105
Shan people 37–8
Shan States 62, 85
Shan States Lines of Communication 99–100
Shwebo 141, 144, 148–51, 154
Shwenyaung 99, 100, 107, 112
Shwenyaung Branch 44, 65, 127
Siam Railways 2
signals and signalling 5, 16–17, 29, 35, 79, 86–7, 110, 123, 127, 142, 147, 148, 151, 152
  absence of (dark territory) 147–57
Simla 169
Simpson, A. F. 48
Singapore 2, 51, 53, 62, 65–6, 70, 187 n.1
  impact on morale and strategic position in Burma by fall of 65–6, 70
Singh, Rampal 127
Sittang Bridge 63–7, 75
Sittang River 65, 68, 69, 90, 99–100
Sittang Valley 89
Sittang Valley State Railway (*see* Rangoon-Mandalay Trunk Line)
6th Army (Chinese) 99
Slim, William (Field Marshal) 16, 86, 89, 125, 131, 134, 164
Soden, James Newton (Col.) 16, 82, 84, 87
Soord, Hector 81–2, 85, 171
South Africa 26
South African War 15
South East Asia Command 162, 169
Southern Railway 18
Southern Shan States Railway (*see* Shwenyaung Branch)
Sri Lanka (Ceylon) 33, 162, 165
Stewart, C. G. (Col.) 151
Stilwell, Joseph (Gen.) 89, 129–30, 163
Stowe, Leland 123
Sung, Brigadier-General 151
Swa 69, 101, 102
Swithinbank, B. 121
Swithinbank Committee 121–2
Syriam 66–7, 113

Tadagale 88
Taikkyi 88–9
Takaw 64
Tantabin Station 152
Tatkon 119–20, 127–8
Taungdwingyi 107, 112–13, 115, 123, 128
Taw-gywe-In 121
technology and technological adaptation (*see* military adaptation and innovation)
Tenasserim 1, 60–7, 71, 73, 80, 88
  withdrawal of Burma Army from 64–5, 70, 73
Thanatpin 68
Thazi-Myingyan-Mandalay Branch Line (*see* Myingyan Branch)
Thazi-Shwenyaung Branch Line (*see* Shwenyaung Branch)
3rd Royal Bombay Sappers and Miners Division 25
Thompson, H. L. (Wing Commander) 53
Thompson, T. O. (Brig.) 16
Thongwa 67–8, 70
Thongwa Branch 67–8
Thongwa Station 67
Tinsukia 175
Tokyo Radio 54

Toungoo 43, 69, 79–84, 89, 97–112, 118, 130
The Transport Act of 1947 18
Transport Co-ordination Board 52–3, 57
Transportation Directorate, India 29, 56, 162
Transportation Directorate, Burma 9, 10, 12, 14, 25–7, 57, 60–1, 73, 75, 78, 82, 107, 118, 169, 171, 178
Transportation Technique 3, 6–8, 10–14, 15–19, 21, 24, 30, 31, 49, 51–2, 60, 71, 74, 77, 99, 103, 134, 156, 157–60, 165, 178, 182–6
Tseng (Brig.) 118, 151
22nd Chinese Division 114, 119, 130
200th Chinese Division 109

Vickers, Wilmot Gordon Hilton 'Phil' (Lt. Gen.) 28, 164, 183, 208 n.12
Vorley, J. S. 152

Wakeford 169
Wakely, Arthur Victor Trocke (Maj. Gen.) 82
Wavell, Archibald (Field Marshal) 68
Weber, Max 43–4, 187 n.10
"webs of empire" 187 n.4 (*see also* knowledge and learning)
Wedgwood, Ralph L. 41–2
Wedgwood Committee 41, 45–7
West Yorkshire Regiment 68
Whapham, C. H. 169
Williams, Captain 127
Williams, F. Seymour (Lt. Col.) 58, 73
Wise, John 138
Woods, Philip 187 n.1
Woolwich 39
World Trade Depression 2–3, 12, 18, 25, 34, 48
Wuntho 150

Yamethin 102, 105–6, 110, 115–18, 120
Ye Branch 44, 62–3
Ye-U 149, 151
Ye-U Branch 44, 135–6, 139, 140
Ye-U Station 149
Yedashe 111, 114–17, 120
Yenangyaung 66, 75, 90, 99, 113
Yezin 119
Yinnyein Station 65
Yitkan 68
Yunnan-Burma Railway 113, 135, 139
Ywadaw 120
Ywataung Junction 43, 84, 92, 135–6, 146, 151

Zeyawadi 110

www.ingramcontent.com/pod-product-compliance
Ingram Content Group UK Ltd.
Pitfield, Milton Keynes, MK11 3LW, UK
UKHW020806290926
15470UKWH00003B/993

9 781350 089457